Ranking Queries on Uncertain Data

ADVANCES IN DATABASE SYSTEMS

Volume 42

Series Editors

Ahmed K. Elmagarmid
Purdue University
West Lafayette, IN 47907

Amit P. Sheth
Wright State University
Dayton, OH 45435

For other titles published in this series, please visit www.springer.com/series/5573

Ming Hua • Jian Pei

Ranking Queries on Uncertain Data

Dr. Ming Hua
Facebook Inc.
S. California Avenue 1601
94304 Palo Alto California
USA
arceehua@fb.com

Dr. Jian Pei
Simon Fraser University
School of Computing Science
University Drive 8888
V5A 1S6 Burnaby British
Columbia
Canada
jpei@cs.sfu.ca

ISSN 1386-2944
ISBN 978-1-4614-2855-8 ISBN 978-1-4419-9380-9 (eBook)
DOI 10.1007/978-1-4419-9380-9
Springer New York Dordrecht Heidelberg London

Printed on acid-free paper

Springer is part of Springer Science+Business Media (www.springer.com)

To my parents {M.H.}
To my wife Jennifer and my daughter
Jacqueline for your love and encouragement
{J.P.}

Preface

"Maturity of mind is the capacity to endure uncertainty."
— John Finley (1935 – 2006)

"Information is the resolution of uncertainty."
— Claude Elwood Shannon (1916 – 2001)

Uncertain data is inherent in many important applications, such as environmental surveillance, market analysis, and quantitative economics research. Due to the importance of those applications and rapidly increasing amounts of uncertain data collected and accumulated, analyzing large collections of uncertain data has become an important task. Ranking queries (also known as top-k queries) are often natural and useful in analyzing uncertain data.

In this monograph, we study the problem of ranking queries on uncertain data. Specifically, we extend the basic uncertain data model in three directions, including uncertain data streams, probabilistic linkages, and probabilistic graphs, to meet various application needs. Moreover, we develop a series of novel ranking queries on uncertain data at different granularity levels, including selecting the most typical instances within an uncertain object, ranking instances and objects among a set of uncertain objects, and ranking the aggregate sets of uncertain objects.

To tackle the challenges on efficiency and scalability, we develop efficient and scalable query evaluation algorithms for the proposed ranking queries. First, we integrate statistical principles and scalable computational techniques to compute exact query results. Second, we develop efficient randomized algorithms to approximate the answers to ranking queries. Third, we propose efficient approximation methods based on the distribution characteristics of query results. A comprehensive empirical study using real and synthetic data sets verifies the effectiveness of the proposed ranking queries and the efficiency of our query evaluation methods.

This monograph can be a reference for academic researchers, graduate students, scientists, and engineers interested in techniques of uncertain data management and analysis, as well as ranking queries. Although the monograph focuses on ranking queries on uncertain data, it does introduce some general principles and models of uncertain data management. Thus, the monograph can also serve as introductory reading approaching the general field of uncertain data management.

Uncertain data processing, management, and exploration in general remain an interesting and fast developing topic in the field of database systems and data analytics. Moreover, ranking queries on uncertain data as a specific topic keeps seeing new progress in both research and engineering development. We believe uncertain data management in general and ranking queries on uncertain data in specific are exciting directions, and still have a huge space for further research. This monograph can inspire some exciting opportunities.

This monograph records the major outcomes of Ming Hua's Ph.D. research at School of Computing Science, Simon Fraser University. This research was an interesting and rewarding journey. We started with several interesting problems concerning effective and efficient queries of massive data, where uncertainty, probability, and typicality play critical roles. It turned out that ranking queries provide a simple and nice way to bind those projects and ideas together. Moreover, we considered various application scenarios, including online analytic style exploration, continuously monitoring of streaming data, data integration, and road network analysis. Only after almost three years we realized that the whole bunch of work can be linked together and weave a nice picture under the theme of ranking queries on uncertain data, as presented in this monograph.

A Ph.D. thesis is never easy. Ming Hua's Ph.D. study is exciting and, at the same time, challenging, for both herself and Jian Pei, the senior supervisor (that is, the thesis advisor). We both remember the sleepless nights before the submission deadlines, the frustration when our submissions were rejected, the suffering moments before some new ideas came to our mind, and the excitement when we obtained breakthroughs afterall. This experience has been casted deeply in our memory forever. We are so lucky to be able to work as a team in those four years.

Acknowledgement

This research would not be possible without the great help and support from many people.

Ming Hua thanks Jian Pei, her senior supervisor and mentor, for his continuous guidance and support during her Ph.D. study at Simon Fraser University. Her gratitude also goes to Funda Ergun, Martin Ester, Lise Getoor, Wei Wang, and Shouzhi Zhang for their insightful comments and suggestions on her research along the way.

Jian Pei thanks Ming Hua for taking the challenge to be his first Ph.D. student at Simon Fraser University. He is deeply grateful to his students at Simon Fraser University. He is always proud of working with those talented students. He is also deeply indebted to his colleagues at Simon Fraser University.

Many ideas in this monograph were resulted from collaboration and discussion with Xuemin Lin, Ada Fu, Ho-fung Leung, and many others. We want to thank our collaborators in the past who have fun together in solving all kinds of data related puzzles. Our gratitude also goes to all anonymous reviewers of our submissions for their invaluable feedback, no matter positive or negative.

We thank Susan Lagerstrom-Fife and Jennifer Maurer at Springer who contributed a lot to the production of this monograph.

This research is supported in part by an NSERC Discovery Grant, an NSERC Discovery Accelerator Supplements Grant, a Simon Fraser University President Research Grant, and a Simon Fraser University Community Trust Endowment Fund. All opinions, findings, conclusions and recommendations in this paper are those of the authors and do not necessarily reflect the views of the funding agencies.

Palo Alto, CA, USA, and Coquitlam, BC, Canada *Ming Hua*
January 2011 *Jian Pei*

Contents

Chapter 1
Introduction

Uncertain data is inherent in several important applications such as sensor network management [1] and data integration [2], due to factors such as data randomness and incompleteness, limitations of measuring equipment and delayed data updates. Because of the importance of those applications and the rapidly increasing amount of uncertain data, analyzing large collections of uncertain data has become an important task.

Ranking queries (also known as top-k queries) [3, 4, 5, 6] are a class of important queries in data analysis. Although ranking queries have been studied extensively in the database research community, uncertainty in data poses unique challenges on the semantics and processing of ranking queries. Traditional queries and evaluation methods on certain data cannot be directly adopted to uncertain data processing. Therefore, practically meaningful ranking queries as well as efficient and scalable query evaluation methods are highly desirable for effective uncertain data analysis.

1.1 Motivation

Recently, there have been an increasing number of studies on uncertain data management and processing [7, 8, 9, 10, 11, 12, 13, 14, 15, 16]. For example, the probabilistic database model [7, 17, 8] and the uncertain object model [18, 19, 20, 21] are developed to describe the uncertainty in data. More details about those models can be found in Chapter 3. In some important application scenarios, various ranking queries can provide intersecting insights into uncertain data.

Example 1.1 (Ranking queries in traffic monitoring applications). Roadside sensors are often used to measure traffic volumes, measure vehicle speeds, or classify vehicles. However, data collected from sensors as such cannot be accurate all the time due to the limitations of equipment and delay or loss in data transfer. Therefore, confidence values are often assigned to such data, based on the specific sensor characteristics, the predicted value, and the physical limitations of the system [22].

Consequently, the sensor readings are inherently uncertain and probabilistic. In this example, we consider three different application scenarios in traffic monitoring.

Scenario 1: Finding the top-k speeding records at a certain time.

Table 1.1 lists a set of synthesized records of vehicle speeds recorded by sensors. Each sensor reports the location, time, and speed of vehicles passing the sensor. In some locations where the traffic is heavy, multiple sensors are deployed to improve the detection quality. Two sensors in the same location (e.g., S206 and S231, as well as S063 and S732 in Table 1.1) may detect the vehicle speed at the same time, such as records R2 and R3, as well as R5 and R6. In such a case, if the speeds reported by multiple sensors are inconsistent, at most one sensor can be correct.

The uncertain data in Table 1.1(a) carries the possible worlds semantics [23, 12, 24, 7] as follows. The data can be viewed as the summary of a set of possible worlds, where a possible world contains a set of tuples governed by some underlying generation rules which constrain the presence of tuples. In Table 1.1, the fact that R2 and R3 cannot be true at the same time can be captured by a generation rule $R2 \oplus R3$. Another generation rule is $R5 \oplus R6$. Table 1.1(b) shows all possible worlds and their existence probability values.

Ranking queries can be used to analyze uncertain traffic records. For example, it is interesting to find out the top-2 speeding records so that actions can be taken to improve the situation. However, in different possible worlds the answers to this question may be different. What a ranking query means on uncertain data in such an application scenario and how to answer a ranking query efficiently are studied in Chapter 5 in this book.

Scenario 2: Monitoring top-k speeding spots in real time.

Table 1.1 contains a set of uncertain records at a certain time. In some applications, a speed sensor will keep sending traffic records to a central server continuously. Therefore, the speeds recorded by each sensor can be modeled as a data stream.

For example, the ARTIMIS center in Cincinnati, Ohio/Kentucky reports the speed, volume and occupancy of road segments every 30 seconds [25]. Table 1.2 is a piece of sample data from ARTIMIS Data Archives[1].

Consider a simple continuous query – continuously reporting a list of top-2 monitoring points in the road network of the fastest vehicle speeds in the last 5 minutes. *One interesting and subtle issue is how we should measure the vehicle speed at a monitoring point. Can we use some simple statistics like the average/median/maximum/minimum speed in the last 5 minutes? Each of such simple statistics may not capture the distribution of the data well. Therefore, new ranking criteria for such uncertain data streams are highly desirable. Moreover, it is important to develop efficient query monitoring algorithms that suit the application need.*

In Chapter 6, we will introduce an uncertain data stream model and a continuous probabilistic threshold top-k query to address this application scenario. Efficient stream specific query evaluation methods will be discussed.

[1] `http://www.its.dot.gov/JPODOCS/REPTS_TE/13767.html`

Record-id	Location	Time	Sensor-id	Speed	Confidence
$R1$	A	07/15/2001 00:01:51	$S101$	60	0.3
$R2$	B	07/15/2001 00:01:51	$S206$	55	0.4
$R3$	B	07/15/2001 00:01:51	$S231$	47	0.5
$R4$	A	07/15/2001 00:01:51	$S101$	45	1.0
$R5$	E	07/15/2001 00:01:51	$S063$	52	0.8
$R6$	E	07/15/2001 00:01:51	$S732$	43	0.2
Generation rules: $R2 \oplus R3$, $R5 \oplus R6$					

(a) Roadside sensor records.

Possible world	Probability	Top-2 speeding records
$W1 = \{R1, R2, R4, R5\}$	0.096	$R1, R2$
$W2 = \{R1, R2, R4, R6\}$	0.024	$R1, R2$
$W3 = \{R1, R3, R4, R5\}$	0.12	$R1, R5$
$W4 = \{R1, R3, R4, R6\}$	0.03	$R1, R3$
$W5 = \{R1, R4, R5\}$	0.024	$R1, R5$
$W6 = \{R1, R4, R6\}$	0.006	$R1, R4$
$W7 = \{R2, R4, R5\}$	0.224	$R2, R5$
$W8 = \{R2, R4, R6\}$	0.056	$R2, R4$
$W9 = \{R3, R4, R5\}$	0.28	$R5, R3$
$W10 = \{R3, R4, R6\}$	0.07	$R3, R4$
$W11 = \{R4, R5\}$	0.056	$R5, R4$
$W12 = \{R4, R6\}$	0.014	$R4, R6$

(b) The possible worlds of Table 1.1(a).

Table 1.1 Speed records reported by sensors.

Scenario 3: Finding optimal paths in road networks with uncertain travel time. *The speed information recorded by sensors, together with the geographic information, can be used to estimate the travel time along each road segment. However, the estimated travel time derived from sensor readings is inherently uncertain and probabilistic, due to the uncertain nature of the collected sensor data. Thus, a road network can be modeled as a simple graph with probabilistic weights.*

Suppose in an uncertain traffic network, from point A to point B there are two paths, P_1 and P_2. The set of travel time samples (in minutes) of P_1 is $\{35, 35, 38, 40\}$ and the set of samples of P_2 is $\{25, 25, 48, 50\}$. Should each sample take a membership probability of 25%, the average travel time on both P_1 and P_2 is 37 minutes. Which path is better?

On the one hand, if a user wants to make sure that the travel time is no more than 40 minutes, path P_1 is better since according to the samples, it has a probability of 100% to meet the travel time constraint, while path P_2 has a probability of only 50% to meet the constraint. On the other hand, if a user wants to go from A to B in 30 minutes, path P_2 should be recommended since the path has a probability of 50% to make it while P_1 has no chance to make it.

In Chapter 8, we will discuss the problem of path queries in probabilistic road networks in detail. ■

# Time	Samp	Speed	Volume	Occupancy
00 : 01 : 51	30	47	575	6
00 : 16 : 51	30	48	503	5
00 : 31 : 51	30	48	503	5
00 : 46 : 51	30	49	421	4
01 : 01 : 52	30	48	274	5
01 : 16 : 52	30	42	275	14
...				

Table 1.2 Data for segment *SEGK*715001 for 07/15/2001 in ARTIMIS Data Archives (Number of Lanes: 4).

Example 1.1 demonstrates the great need for ranking queries in uncertain data analysis. In traditional data analysis for deterministic data, ranking queries play an important role by selecting the subset of records of interest according to user specified criteria. With the rapidly increasing amount of uncertain data, ranking queries have become even more important, since the uncertainty in data not only increases the scale of data but also introduces more difficulties in understanding and analyzing the data.

1.2 Challenges

While being useful in many important applications, ranking queries on uncertain data pose grand challenges to query semantics and processing.

Challenge 1 What are the uncertain data models that we need to adopt?
Example 1.1 illustrates three different application scenarios in ranking the information obtained from traffic sensors. This not only shows the great use of ranking queries on uncertain data, but also raises a fundamental question: *how can we develop uncertain data models that capture the characteristics of data and suit application needs*?

In particular, we need to consider the following three aspects. First, is the uncertain data static or dynamic? Second, how to describe the dependencies among uncertain data objects? Third, how can we handle complex uncertain data like a graph?

Challenge 2 How to formulate probabilistic ranking queries?
As shown in Example 1.1, different ranking queries on uncertain data can be asked according to different application needs. In Scenario 1, we want to select the records ranked in top-k with high confidence, while in Scenario 2, the objective is to find the sensors whose records are ranked in top-k with probabilities no smaller than a threshold in a time window. Last, in Scenario 3, we are interested in finding paths such that the sums of the (uncertain) travel time along the path are ranked at the top.

Therefore, it is important to develop meaningful ranking queries according to different application interests. Moreover, the probability associated with each data record introduces a new dimension in ranking queries. How to leverage the probabilities in ranking queries remains challenging in uncertain data analysis.

Challenge 3 How to develop efficient and scalable query processing methods? Evaluating ranking queries on uncertain data is challenging. On the one hand, traditional ranking query processing methods cannot be directly applied since they do not consider how to handle probabilities. On the other hand, although some standard statistical methods such as Bayesian Statistics [26] can be applied to analyzing uncertain data in some applications, efficiency and scalability issues are usually not well addressed.

Meanwhile, as shown in Example 1.1, uncertain data is a summary of all possible worlds. Therefore, a naïve way to answer a ranking query on uncertain data is to apply the query to all possible worlds and summarize the answers to the query. However, it is often computationally prohibitive to enumerate all possible worlds. Thus, we need to develop efficient and scalable query evaluation methods for ranking queries on uncertain data.

1.3 Focus of the Book

In this book, we discuss probabilistic ranking queries on uncertain data and address the three challenges in Section 1.2. Specially, we focus on the following aspects.

- **We introduce three extended uncertain data models.**
  To address Challenge 1, we first study two basic uncertain data models, the *probabilistic database model* and the *uncertain object model*, and show that the two models are equivalent.
  Then, we develop three extended uncertain object model, to address three important application scenarios. The first extension, the *uncertain data stream model*, describes uncertain objects whose distributions evolve over time. The second extension, the *probabilistic linkage model*, introduces inter-object dependencies into uncertain objects. The third extension, the *uncertain road network model*, models the weight of each edge in road networks as an uncertain object.
- **We discuss five novel problems of ranking uncertain data.**
  To address Challenge 2, we formulate five novel ranking problems on uncertain data models from multiple aspects and levels.
  First, from the *data granularity* point of view, we study the problems of ranking instances within a single uncertain object, ranking instances among multiple uncertain objects, ranking uncertain objects and ranking the aggregates of a set of uncertain objects. Second, from the *ranking scope* point of view, we study ranking queries within an uncertain object and among multiple uncertain objects. Third, from the query type point of view, we discuss two categories of ranking queries considering both ranking criteria and the probability constraint.
- **We discuss three categories of query processing methods.**

To address Challenge 3, we develop three categories of efficient query answering methods for each of the proposed ranking queries on uncertain data.
First, we integrate statistical principles and scalable computational techniques to compute the exact answers to queries. Second, we develop efficient randomized algorithms to estimate the answers to ranking queries. Third, we devise efficient approximation methods based on the distribution characteristics of answers to ranking queries.

A software package including the implementation of our algorithms and the probabilistic data generator was released to public in March, 2008, and is downloadable at `http://www.cs.sfu.ca/~jpei/Software/PTKLib.rar`.

1.4 Organization of the Book

The remainder of the book is organized as follows.

- In Chapter 2, we formulate the uncertain data models and probabilistic ranking queries that will be studied in this book.
- Chapter 3 reviews the related literature on uncertain data processing and principles of statistics and probability theory that will be used.
- In Chapter 4, we introduce the top-k typicality queries on uncertain data, which find the top-k most typical instances for an uncertain object. We answer two fundamental questions. First, given an uncertain object with a large number of instances, how can we model the typicality of each instance? Second, how to efficiently and effectively select the most representative instances for the object? This is essentially the problem of ranking instances within an uncertain object.
- In Chapter 5, we discuss probabilistic ranking queries on probabilistic databases, which select the instances in different uncertain objects whose probabilities of being ranked top-k are high. Although it is an extension of the problem in Chapter 4, the query evaluation techniques are significantly different.
- We extend probabilistic ranking queries from static data to dynamic data in Chapter 6. The objective is to continuously report the answers to a probabilistic ranking query on a set of uncertain data streams, as illustrated in the second application scenario in Example 1.1. We develop stream-specific query evaluation methods that are highly space efficient.
- In Chapter 7, we introduce inter-object dependencies among uncertain data object and study probabilistic ranking queries on a set of dependent uncertain objects, which can find important applications in data integration. We show that the model is a special case of Markov Random Fields. Moreover, we develop efficient methods to evaluate ranking queries on the proposed uncertain data model.
- In Chapter 8, we extend the probabilistic ranking queries to uncertain road networks, where the weights of each edge in the network is an uncertain object. We want to select the paths having high confidences of being ranked top-k in terms

of shortest travel time. We introduce several interesting path queries and discuss the efficient query evaluation.

- In Chapter 9, we discuss several interesting extensions and applications of probabilistic ranking queries on uncertain data, and present some future research directions.

Chapter 2
Probabilistic Ranking Queries on Uncertain Data

In this chapter, we formulate the probabilistic ranking queries on uncertain data. We first introduce two basic uncertain data models and basic probabilistic ranking queries. Then, we discuss three extended uncertain data models that suit different application scenarios. Ranking queries on the extended uncertain data models are also developed.

Frequently used definitions and notations are listed in Table 2.1.

2.1 Basic Uncertain Data Models

We consider uncertain data in the *possible worlds* semantics model [23, 12, 24, 7], which has been extensively adopted by the recent studies on uncertain data processing, such as [17, 8, 21]. Technically, uncertain data can be represented in two ways.

2.1.1 Uncertain Object Model

An uncertain object O [18, 19, 20, 21] is conceptually governed by an underlying random variable X. Theoretically, if X is a continuous random variable, the distribution of X can be captured by a *probability density function* (*PDF* for short); if X is a discrete random variable, its distribution can be described by a *probability mass function* (*PMF* for short). In practice, the PDF or PMF of a random variable is often unavailable. Instead, a sample set of instances $x_1, \cdots, x_m$ are used to approximate the distribution of X, where each instance takes a membership probability. For an instance $x_i \in X$ $(1 \leq i \leq m)$, the membership probability of x_i measures the likelihood that x_i will occur. Due to the unavailability of X's PDF or PMF, in this book, we represent an uncertain object O using the set of samples $x_1, \cdots, x_m$ generated by the underlying random variable.

Notation	Description
$O=\{o_1,\cdots,o_m\}$	an uncertain object contains m instances
$\mathscr{O}$	a set of uncertain objects
$T=\{t_1,\cdots,t_n\}$	a table with n tuples
$R: t_{r_1}\oplus\cdots\oplus t_{r_m}$	a generation rule specifying the exclusiveness among $t_{r_1},\cdots,t_{r_m}$
$\mathscr{R}$	a set of generation rules
W	a possible world
$\mathscr{W}$	a set of possible worlds
$O=o_1,o_2,\cdots$	an uncertain data stream
$W^t_\omega(\mathscr{O})$	a set of uncertain data streams in sliding window W^t_ω
$\mathscr{L}(t_A,t_B)$	a probabilistic linkage between tuples t_A and t_B
$G(V,E,W)$	a simple graph with probabilistic weights W

Table 2.1 Summary of definitions and frequently used notations.

Definition 2.1 (Uncertain object). An **uncertain object** is a set of instances $O=\{o_1,\cdots,o_m\}$ such that each instance o_i $(1\leq i\leq m)$ takes a **membership probability** $Pr(o_i)>0$, and $\sum_{i=1}^m Pr(o_i)=1$. ■

The **cardinality** of an uncertain object $O=\{o_1,\cdots,o_m\}$, denoted by $|O|$, is the number of instances contained in O. We denote the set of all uncertain objects as $\mathscr{O}$.

2.1.1.1 Possible Worlds Semantics

In the basic uncertain object model, we assume that the distributions of uncertain objects are independent from each other. Correlations among uncertain objects are discussed in Section 2.3.2. The uncertain objects carry the possible worlds semantics.

Definition 2.2 (Possible worlds of uncertain objects). Let $\mathscr{O}=\{O_1,\cdots,O_n\}$ be a set of uncertain objects. A **possible world** $W=\{o_1,\cdots,o_n\}$ $(o_i\in O_i)$ is a set of instances such that one instance is taken from each uncertain object. The **existence probability** of W is $Pr(W)=\prod_{i=1}^{n} Pr(o_i)$. ■

Let $\mathscr{W}$ denote the set of all possible worlds, we have the following property.

Corollary 2.1 (Number of possible worlds). *For a set of uncertain objects $\mathscr{O}=\{O_1,\ldots,O_n\}$, let $|O_i|$ be the cardinality of object O_i $(1\leq i\leq n)$, the number of all possible worlds is*

$$|\mathscr{W}|=\prod_{i=1}^{n}|O_i|.$$

Moreover,

$$Pr(\mathscr{W})=\sum_{w\in\mathscr{W}} Pr(w)=1$$

■

Example 2.1 (Uncertain objects). Table 1.2 is an example of a uncertain object with 6 instances. Each instance takes an equal membership probability $\frac{1}{6}$. ∎

2.1.2 Probabilistic Database Model

In some other studies, the probabilistic database model is used to represent uncertain data. A probabilistic database [17] is a finite set of probabilistic tables defined as follows.

Definition 2.3 (Probabilistic table). A **probabilistic table** contains a set of uncertain tuples T and a set of generation rules $\mathscr{R}$. Each **uncertain tuple** $t \in T$ is associated with a **membership probability** $Pr(t) > 0$. Each **generation rule** (or **rule** for short) $R \in \mathscr{R}$ specifies a set of exclusive tuples in the form of $R : t_{r_1} \oplus \cdots \oplus t_{r_m}$ where $t_{r_i} \in T$ $(1 \leq i \leq m)$, $Pr(t_{r_i} \wedge t_{r_j}) = 0$ $(1 \leq i, j \leq m, i \neq j)$ and $\sum_{i=1}^{m} Pr(t_{r_i}) \leq 1$. ∎

The probabilistic database model also follows the possible worlds semantics. The generation rule R constrains that, among all tuples $t_{r_1}, \cdots, t_{r_m}$ involved in the rule, at most one tuple can appear in a possible world. R is a *singleton rule* if there is only one tuple involved in the rule, otherwise, R is a *multi-tuple rule*. The **cardinality** of a generation rule R, denoted by $|R|$, is the number of tuples involved in R.

Definition 2.4 (Possible worlds of a probabilistic table). Given a probabilistic table T, a *possible world* W is a subset of T such that for each generation rule $R \in \mathscr{R}_T$, $|R \cap W| = 1$ if $Pr(R) = 1$, and $|R \cap W| \leq 1$ if $Pr(R) < 1$. The existence probability of W is

$$Pr(W) = \prod_{R \in \mathscr{R}_T, |R \cap W| = 1} Pr(R \cap W) \prod_{R \in \mathscr{R}_T, R \cap W = \emptyset} (1 - Pr(R))$$

∎

Corollary 2.2 (Number of possible worlds). *For an uncertain table T with a set of generation rules $\mathscr{R}_T$, the number of all possible worlds is*

$$|\mathscr{W}| = \prod_{R \in \mathscr{R}_T, Pr(R) = 1} |R| \prod_{R \in \mathscr{R}_T, Pr(R) < 1} (|R| + 1)$$

∎

Example 2.2 (Probabilistic tables). Table 1.1(a) is an example of a probabilistic table with 6 uncertain tuples and 2 multi-tuple generation rules $R2 \oplus R3$ and $R5 \oplus R6$. The corresponding possible worlds are shown in Table 1.1(b). ∎

2.1.3 Converting Between the Uncertain Object Model and the Probabilistic Database Model

Interestingly, the uncertain object model and the probabilistic database model are equivalent.

- **Converting from the uncertain object model to the probabilistic database model.** A set of uncertain objects can be represented by a probabilistic table as follows. For each instance o of an uncertain object O, we create a tuple t_o, whose membership probability is $f(o)$. For each uncertain object $O = \{o_1, \cdots, o_m\}$, we create one generation rule $R_O = t_{o_1} \oplus \cdots \oplus t_{o_m}$.
- **Converting from the probabilistic database model to the uncertain object model.** A probabilistic table can be represented by a set of uncertain objects with discrete instances. For each tuple t in a probabilistic table, we create an instance o_t, whose probability mass function is $f(o_t) = Pr(t)$. For a generation rule $R : t_{r_1} \oplus \cdots \oplus t_{r_m}$, we create an uncertain object O_R, which includes instances $o_{t_{o_1}}, \cdots, o_{t_{om}}$ corresponding to $t_{r_1}, \cdots, t_{r_m}$, respectively. Moreover, if $\sum_{i=1}^{m} Pr(t_{r_i}) < 1$, we create another instance $o_\emptyset$ whose probability mass function is $f(o_\emptyset) = 1 - \sum_{i=1}^{m} Pr(t_{r_i})$, and add $u_\emptyset$ to the uncertain object O_R.

Example 2.3 (Converting between two models). $R1$ in Table 1.1(a) can be converted to an uncertain object $O_1 = \{R1, \neg R1\}$ where $Pr(R1) = 0.3$ and $Pr(\neg R1) = 0.7$. Moreover, generation rule $R2 \oplus R3$ in Table 1.1(a) can be converted to uncertain object $O_{1,2} = \{R2, R3, \neg R23\}$ where $Pr(R2) = 0.4$, $Pr(R3) = 0.5$ and $Pr(\neg R23) = 0.1$. ■

2.2 Basic Ranking Queries on Uncertain Data

In this section, we discuss various types of ranking queries on the uncertain object model. Since the uncertain object model and the probabilistic database model are equivalent, the queries discussed in this section can also be applied to the probabilistic database model.

Depending on different application scenarios, probabilistic ranking queries can be applied at one of the three *granularity levels.*

- The *instance* probabilistic ranking queries return the instances satisfying query conditions. We develop two classes of instance probabilistic ranking queries. The first are *top-k typicality queries*, which rank instances in an uncertain object according to how typical each instance is. The second are *probabilistic ranking queries*, which rank instances in multiple objects according to the probability that each instance is ranked top-k. The two classes of queries will be discussed in Sections 2.2.1 and 2.2.2, respectively.
- The *object* probabilistic ranking queries find the object satisfying query conditions, which will be discussed in Section 2.2.3.

- The *object set* probabilistic ranking queries apply the query condition to each object set and return the object set that satisfy the query. We defer the discussion on ranking uncertain object sets to Section 2.3.3 in the context of uncertain road networks.

2.2.1 Ranking Instances in An Uncertain Object

Given an uncertain object with a large number of instances that are samples taken from an underlying random variable, how can we understand and analyze this object? An effective way is to find the most typical instances among all instances of the uncertain object. We develop a class of top-k typicality queries which can serve for this purpose.

Example 2.4 (Top-k typicality queries).

Jeff is a junior basketball player who dreams to play in the NBA. As the NBA has more than 400 *active players, they are quite diverse. Jeff may want to know some representative examples of NBA players. Top-k typicality queries can help.*

We can model the group of NBA players as an uncertain object in the space of technical statistics, which can be described by a likelihood function. Each player is an instance of the uncertain object.

- ***Top-k simple typicality queries.***
  Jeff asks, "Who are the top-3 most typical NBA players?" *The player who has the maximum likelihood of being NBA players is the most typical. This leads to our first typicality measure – the* simple typicality. *A top-k simple typicality query finds the k most typical instances in an uncertain object.*
- ***Top-k discriminative typicality queries.***
  Jeff is particularly interested in becoming a guard. "Who are the top-3 most typical guards distinguishing guards from other players?" *Simple typicality on the set of guards is insufficient to answer the question, since it is possible that a typical guard may also be typical among other players. Instead, players that are typical among all guards but are not typical among all non-guard players should be found.*
  In order to address this demand, we can model the group of guards as a target uncertain object O_g and the set of other players as the other uncertain object O. The notion of discriminative typicality *measures how an instance is typical in one object but not typical in the other object. Given two uncertain objects O and S, let O be the target object, a top-k discriminative typicality query finds the k instances with the highest discriminative typicality values in O.*
- ***Top-k representative typicality queries.***
  NBA guards may still have some sub-groups. For example, the fresh guards and the experienced guards, as well as the shooting guards and the point guards. Jeff wants to learn different types of guards, without a clear idea about what types

there are. So he asks, "Who are the top-3 typical guards in whole representing different types of guards?"

Simple typicality does not provide the correct answer to this question, since the 3 players with the greatest simple typicality may be quite similar to each other, while some other popular players different from those three may be missed. Discriminative typicality does not help either, because the exact types of guards and their members are unknown.

To solve this problem, we develop the notion of representative typicality *that measures how an instance is typical in an uncertain object different from the already reported typical instances. Given an uncertain object O, a top-k representative typicality query finds a set of k instances of O with the highest representative typicality scores.* ■

By default, we consider an uncertain object O on attributes $A_1, \cdots, A_n$. Let $A_{i_1}, \cdots, A_{i_l}$ be the attributes on which the typicality queries are applied ($1 \leq i_j \leq n$ for $1 \leq j \leq l$) and $d_{A_{i_1}, \cdots, A_{i_l}}(x, y)$ be the distance between two instances x and y in S on attributes $A_{i_1}, \cdots, A_{i_l}$. When $A_{i_1}, \cdots, A_{i_l}$ are clear from context, $d_{A_{i_1}, \cdots, A_{i_l}}(x, y)$ is abbreviated to $d(x, y)$.

We address the top-k typicality problem in a generic metric space. Therefore, the distance metric d should satisfy the triangle inequality.

2.2.1.1 Simple Typicality

By intuition and as also suggested by the previous research in psychology and cognitive science (as will be reviewed in Section 3.3.1), an instance o in O is more typical than the others if o is more likely to appear in O. As discussed in Section 2.1.1, the set of instances in O on attributes $A_1, \cdots, A_n$ can be viewed as a set of independent and identically distributed samples of an n-dimensional random vector $\mathscr{X}$ that takes values in the Cartesian product space $D = D_{A_1} \times \cdots \times D_{A_n}$, where D_{A_i} is the domain of attribute A_i ($1 \leq i \leq n$). The likelihood of $o \in O$, given that o is a sample of $\mathscr{X}$, can be used to measure the typicality of o.

Definition 2.5 (Simple typicality). Given an uncertain object O on attributes $A_1, \cdots, A_n$ and a subset of attributes $A_{i_1}, \cdots, A_{i_l}$ ($1 \leq i_j \leq n$ for $1 \leq j \leq l$) of interest, let $\mathscr{X}$ be the n-dimensional random vector generating the instances in O, the **simple typicality** of an instance $o \in O$ with respect to $\mathscr{X}$ on attributes $A_{i_1}, \cdots, A_{i_l}$ is defined as $T_{A_{i_1}, \cdots, A_{i_l}}(o, \mathscr{X}) = L_{A_{i_1}, \cdots, A_{i_l}}(o|\mathscr{X})$ where $L_{A_{i_1}, \cdots, A_{i_l}}(o|\mathscr{X})$ is the likelihood [27] of o on attributes $A_{i_1}, \cdots, A_{i_l}$, given that o is a sample of $\mathscr{S}$. ■

In practice, since the distribution of random vector $\mathscr{X}$ is often unknown, we use $T_{A_{i_1}, \cdots, A_{i_l}}(o, O) = L_{A_{i_1}, \cdots, A_{i_l}}(o|O)$ as an estimator of $T_{A_{i_1}, \cdots, A_{i_l}}(o, \mathscr{X})$, where $L_{A_{i_1}, \cdots, A_{i_l}}(o|O)$ is the posterior probability of an object o on attributes $A_{i_1}, \cdots, A_{i_l}$ given O [27].

$L_{A_{i_1}, \cdots, A_{i_l}}(o|O)$ can be computed using density estimation methods. We adopt the commonly used kernel density estimation method, which does not require any distribution assumption on O. The general idea is to use a kernel function to approximate

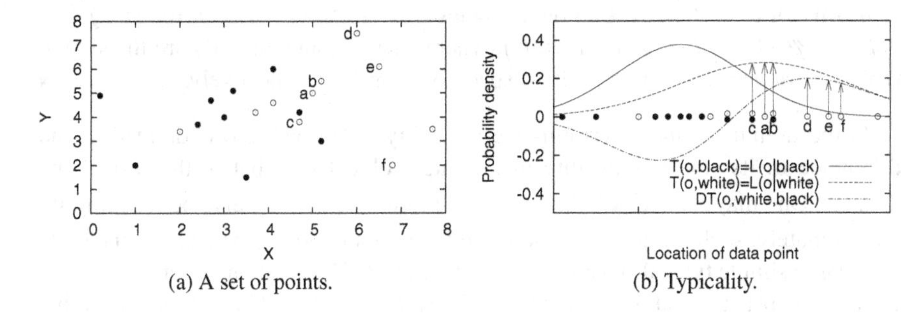

(a) A set of points. (b) Typicality.

Fig. 2.1 The simple typicality and discriminative typicality curves of a set of points.

the probability density around each observed sample. More details will be discussed in Chapter 4.

Hereafter, unless specified otherwise, the simple typicality measure refers to the estimator $T_{A_{i_1},\cdots,A_{i_l}}(o,O)$. Moreover, for the sake of simplicity, when $A_{i_1},\cdots,A_{i_l}$ are clear from context, $T_{A_{i_1},\cdots,A_{i_l}}(o,O)$ and $L_{A_{i_1},\cdots,A_{i_l}}(o|O)$ are abbreviated to $T(o,O)$ and $L(o|O)$, respectively.

Given an uncertain object O on attributes $A_{i_1},\cdots,A_{i_l}$ of interest, a predicate P and a positive integer k, a **top-k simple typicality query** returns, from the set of instances in O satisfying predicate P, the k instances having the largest simple typicality values that are computed on attributes $A_{i_1},\ldots,A_{i_l}$.

Example 2.5 (Top-k simple typicality queries). Consider the set of points belong to an uncertain object in Figure 2.1(a). A top-3 simple typicality query on attribute X with predicate COLOR = white returns the 3 white points having the largest simple typicality values computed on attribute X.

Figure 2.1(b) projects the points in T to attribute X. The likelihood function of the white points and that of the black points on attribute X are labeled as $L(o|white)$ and $L(o|black)$ in the figure, respectively, while we will discuss how to compute the likelihood values in Chapter 4. Points a, b and c have the highest likelihood values among all white points, and thus should be returned as the answer to the query. ■

2.2.1.2 Discriminative Typicality

Given two uncertain objects O and S, which instance is the most typical in O but not in S? We use the discriminative typicality to answer such a question. By intuition, an instance $o \in O$ is typical and discriminative in O if the difference between its typicality in O and that in S is large.

Definition 2.6 (Discriminative typicality). Given two uncertain objects O and S on attributes $A_1,\cdots,A_n$ (O is the **target object**), let $\mathscr{U}$ and $\mathscr{V}$ be the n-dimensional random vectors generating the instances in O and S, respectively, the **discriminative**

typicality of an instance $o \in O$ on attributes $A_{i_1}, \cdots, A_{i_l}$ $(1 \leq i_j \leq n$ for $1 \leq j \leq l)$ is $DT(o, \mathcal{U}, \mathcal{V}) = T(o, \mathcal{U}) - T(o, \mathcal{V})$, where $T(o, \mathcal{U})$ and $T(o, \mathcal{V})$ are the simple typicality values of instance o with respect to $\mathcal{U}$ and $\mathcal{V}$, respectively. ■

In the definition, the discriminative typicality of an instance is defined as the difference of its simple typicality in the target object and that in the rest of the data set. One may wonder whether using the ratio $\frac{T(o,\mathcal{U})}{T(o,\mathcal{V})}$ may also be meaningful. Unfortunately, such a ratio-based definition may not choose a typical instance that has a large simple typicality value with respect to $\mathcal{U}$. Consider an extreme example. Let o be an instance that is very atypical with respect to $\mathcal{U}$ and has a typicality value of nearly 0 with respect to $\mathcal{V}$. Then, o still has an infinite ratio $\frac{T(o,\mathcal{U})}{T(o,\mathcal{V})}$. Although o is discriminative between $\mathcal{U}$ and $\mathcal{V}$, it is not typical with respect to $\mathcal{U}$ at all.

Due to the unknown distribution of random vectors $\mathcal{U}$ and $\mathcal{V}$, we use $DT(o, O, S) = T(o, O) - T(o, S)$ to estimate $DT(o, \mathcal{U}, \mathcal{V})$, where $T(o, O)$ and $T(o, S)$ are the estimators of $T(o, \mathcal{U})$ and $T(o, \mathcal{V})$, respectively.

Given a set of uncertain instances on attributes $A_{i_1}, \ldots, A_{i_l}$ of interest, a predicate P and a positive integer k, a **top-k discriminative typicality query** treats the set of instances satisfying P as the target object, and returns the k instances in the target object having the largest discriminative typicality values computed on attributes $A_{i_1}, \ldots, A_{i_l}$.

Example 2.6 (Top-k discriminative typicality queries). Consider the set of points in Figure 2.1(a) again and a top-3 discriminative typicality query on attribute X with predicate COLOR = white.

The discriminative typicality $DT(o, white, black)$ for each instance $o \in white$ is plotted in the figure, where white and black denote the two uncertain objects, the one with white points as instances and the one with black points as instances, respectively. To see the difference between discriminative typicality and simple typicality, consider instance a, b and c, which have large simple typicality values among all white points. However, they also have relatively high simple typicality values as a member in the subset of black points comparing to other white points. Therefore, they are not discriminative. Points $\{d, e, f\}$ are the answer to the query, since they are discriminative. ■

2.2.1.3 Representative Typicality

The answer to a top-k simple typicality query may contain some similar instances, since the instances with similar attribute values may have similar simple typicality scores. However, in some situations, it is redundant to report many similar instances. Instead, a user may want to explore the uncertain object by viewing typical instances that are different from each other but jointly represent the uncertain object well.

Suppose a subset of instances $A \subset O$ is chosen to represent O. Each instances in $(O - A)$ is best represented by the closest instance in A. For each $o \in A$, we define the representing region of o.

Definition 2.7 (Representing region). Given an uncertain object O on attributes $A_1, \cdots, A_n$ and a subset of instances $A \subset O$, let $D = D_{A_1} \times \cdots \times D_{A_n}$ where D_{A_i} is the domain of attribute A_i ($1 \leq i \leq n$), the **representing region** of an instance $o \in A$ is $D(o,A) = \{x | x \in D, d(x,o) = \min_{y \in A} d(x,y)\}$, where $d(x,y)$ is the distance between objects x and y. ■

To make A representative as a whole, the representing region of each instance o in A should be fairly large and o should be typical in its own representing region.

Definition 2.8 (Group typicality). Given an uncertain object O on attributes $A_1, \cdots, A_n$ and a subset of instances $A \subset O$, let $\mathscr{X}$ be the n-dimensional random vector generating the instances in O, the **group typicality** of A on attributes $A_{i_1}, \cdots, A_{i_l}$ ($1 \leq i_j \leq n$, $1 \leq j \leq l$) is $GT(A, \mathscr{X}) = \sum_{o \in A} T(o, \mathscr{X}_{D(o,A)}) \cdot Pr(D(o,A))$, where $T(o, \mathscr{X}_{D(o,A)})$ is the simple typicality of o with respect to $\mathscr{X}$ in o's representing region $D(o,A)$ and $Pr(D(o,A))$ is the probability of $D(o,A)$. ■

Since the distribution of $\mathscr{X}$ is unknown, we can estimate the group typicality $GT(A, \mathscr{X})$ as follows. For any instance $o \in A$, let $N(o,A,O) = \{x | x \in O \cap D(o,A)\}$ be the set of instances in O that lie in $D(o,A)$, $Pr(D(o,A))$ can be estimated using $\frac{|N(o,A,O)|}{|O|}$. The group typicality $GT(A, \mathscr{X})$ is estimated by $GT(A,O) = \sum_{o \in A} T(o, N(o,A,O)) \cdot \frac{|N(o,A,O)|}{|O|}$, where $T(o, N(o,A,O))$ is the estimator of simple typicality $T(o, \mathscr{X}_{D(o,A)})$, since $N(o,A,O)$ can be viewed as a set of independent and identically distributed samples of $\mathscr{X}$ that lie in $D(o,A)$.

The group typicality score measures how representative a group of instances is. The *size-k most typical group problem* is to find k instances as a group such that the group has the maximum group typicality. Unfortunately, the problem is NP-hard, since it has the discrete k-median problem as a special case, which was shown to be NP-hard [28].

Moreover, top-k queries are generally expected to have the monotonicity in answer sets. That is, the result of a top-k query is contained in the result of a top-k' query where $k < k'$. However, an instance in the most typical group of size k may not be in the most typical group of size k' ($k < k'$). For example, in the data set illustrated in Figure 2.2, the size-1 most typical group is $\{A\}$ and the size-2 most typical group is $\{B,C\}$, which does not contain the size-1 most typical group. Therefore, the size-k most typical group is not suitable to define the top-k representative typicality.

To enforce monotonicity, we adopt a greedy approach.

Definition 2.9 (Representative typicality). Given an uncertain object O and a **reported answer set** $A \subset O$, let $\mathscr{X}$ be the random vector with respect to instances in O, the **representative typicality** of an instance $o \in (O - A)$ is $RT(o,A,\mathscr{X}) = GT(A \cup \{o\}, \mathscr{X}) - GT(A, \mathscr{X})$, where $GT(A \cup \{o\}, \mathscr{X})$ and $GT(A, \mathscr{X})$ are the group typicality values of subsets $A \cup \{o\}$ and A, respectively. ■

In practice, we use $RT(o,A,O) = GT(A \cup \{o\}, O) - GT(A,O)$ to estimate $RT(o,A,\mathscr{X})$, where $GT(A,O)$ and $GT(A \cup \{o\}, O)$ are the estimators of $GT(A, \mathscr{X})$ and $GT(A \cup \{o\}, \mathscr{X})$, respectively.

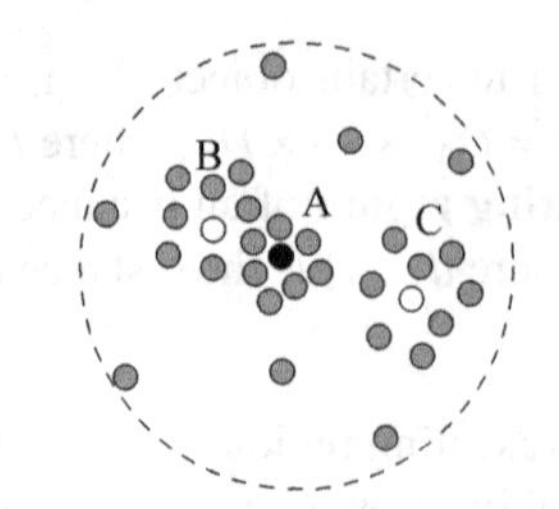

Fig. 2.2 Non-monotonicity of size-k most typical group.

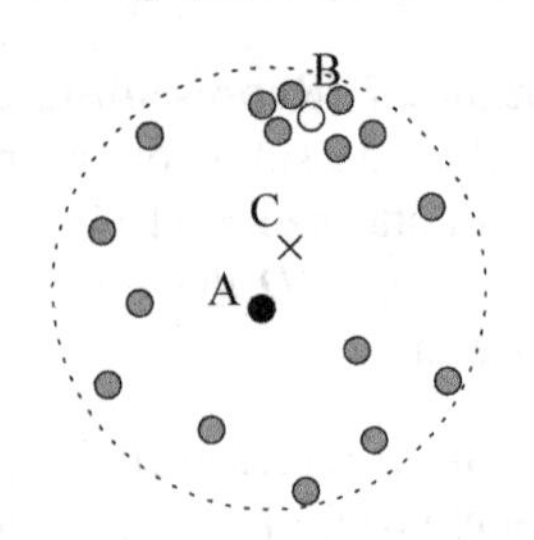

Fig. 2.3 Medians, means and typical objects.

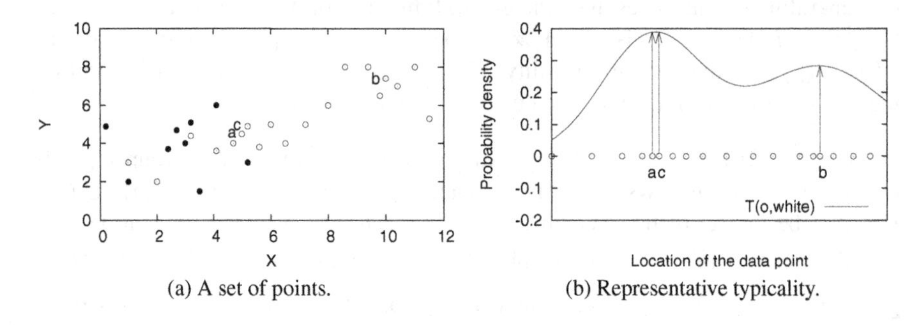

(a) A set of points.

(b) Representative typicality.

Fig. 2.4 The answer to a top-2 representative typicality query on a set of points.

Given an uncertain object O on attributes $A_{i_1},\ldots,A_{i_l}$ of interest, a predicate P and a positive integer k, a **top-k representative typicality query** returns k instances $o_1,\ldots,o_k$ from the set of instances in O satisfying predicate P, such that o_1 is the instance having the largest simple typicality, and, for $i > 1$,

$$o_i = \arg \max_{o \in O - \{o_1,\ldots,o_{i-1}\}} RT(o, \{o_1,\ldots,o_{i-1}\}, O).$$

The representative typicality values are computed on attributes $A_{i_1},\ldots,A_{i_l}$.

Example 2.7 (Top-k representative typicality queries). Consider the set of points in Figure 2.4(a) and a top-2 representative typicality query on attribute X with predicate COLOR = white.

We project the white points to attribute X and plot the simple typicality scores of the white points, as shown in Figure 2.4(b). Points a and c have the highest simple typicality scores. However, if we only report a and c, then the dense region around a is reported twice, but the dense region around b is missed. A top-2 representative typicality query will return a and b as the answer. ■

2.2.2 Ranking Uncertain Instances in Multiple Uncertain Objects

Given multiple uncertain objects, how can we select a small subset of instances meeting users' interests? Ranking queries (also known as top-k queries) [3, 4, 5, 6] are a class of important queries in data analysis that allows us to select the instances ranked top according to certain user specified scoring functions. We consider the top-k selection query model [29].

Definition 2.10 (Top-k selection queries). For a set of instances S, each instance $o \in S$ is associated with a set of attributes A. Given a predicate P on A, a ranking function $f : S \rightarrow R$ and a integer $k > 0$, a **top-k selection query** $Q^k_{P,f}$ returns a set of instances $Q^k_{P,f}(S) \subseteq S_P$, where S_P is the set of instances satisfying P, $|Q^k_{P,f}(S)| = \min\{k, |S_P|\}$ and $f(o) > f(o')$ for any instances $o \in Q^k_{P,f}(S)$ and $o' \in S_P - Q^k_{P,f}(S)$.■

To keep our presentation simple, we assume that the top-k selection queries in our discussion select all instances in question. That is $S_P = S$. Those selection predicates can be implemented efficiently as filters before our ranking algorithms are applied. Moreover, we assume that the ranking function f in a top-k selection query can be efficiently applied to an instance o to generate a score $f(o)$. When it is clear from context, we also write $Q^k_{P,f}$ as Q^k for the sake of simplicity.

2.2.2.1 Ranking Probabilities

How can we apply a top-k selection query to a set of uncertain objects? Since each object appears as a set of instances, we have to rank the instances in the possible worlds semantics.

A top-k selection query can be directly applied to a possible world that consists of a set of instances. In a possible word, a top-k selection query returns k instances. We define the rank-k probability and top-k probability for instances and objects as follows.

Given a set of uncertain objects and a *ranking function* f, all instances of the uncertain objects can be ranked according to the ranking function. For instances o_1 and o_2, $o_1 \preceq_f o_2$ if o_1 is ranked higher than or equal to o_2 according to f. The *ranking order* $\preceq_f$ is a total order on all instances.

Definition 2.11 (rank-k Probability and top-k probability). For an instance o, the **rank-k probability** $Pr(o,k)$ is the probability that o is ranked at the k-th position in possible worlds according to f, that is

$$Pr(o,k) = \sum_{W \in \mathscr{W} \text{ s.t. } o=W_f(k)} Pr(W) \tag{2.1}$$

where $W_f(k)$ denotes the instance ranked at the k-th position in W.

The **top-k probability** $Pr^k(o)$ is the probability that o is ranked top-k in possible worlds according to f, that is,

$$Pr^k(o) = \sum_{j=1}^{k} Pr(o,j). \quad (2.2)$$

■

2.2.2.2 Ranking Criteria

Given a *rank parameter* $k > 0$ and a probability threshold $p \in (0,1]$, a *probability threshold top-k query* (*PT-k query* for short) [30, 31] finds the instances whose top-k probabilities are no less than p.

Definition 2.12 (PT-k query and top-(k,l) query). Given a *rank parameter* $k > 0$ and a probability threshold $p \in (0,1]$, a **probabilistic threshold top-k query** (*PT-k query* for short) [30, 31] finds the instances whose top-k probabilities are no less than p.

Alternatively, a user can use an *answer set size constraint* $l > 0$ to replace the probability threshold p and issue a **top-(k,l) query** [32, 33], which finds the top-l tuples with the highest top-k probabilities. ■

Now, let us consider the reverse queries of PT-k queries and top-(k,l) queries.

For an instance o, given a probability threshold $p \in (0,1]$, the *p-rank* of o is the minimum k such that $Pr^k(o) \geq p$, denoted by $MR_p(o) = \min\{k | Pr^k(o) \geq p\}$.

Definition 2.13 (RT-k query and top-(p,l) query). Given a *probability threshold* $p \in (0,1]$ and a *rank threshold* $k > 0$, a **rank threshold query** (*RT-k query* for short) to retrieve the instances whose p-ranks are at most k. RT-k queries are reverse queries of PT-k queries.

Alternatively, a user can replace the rank threshold by an *answer set size constraint* $l > 0$ and issue a **top-(p,l) query**, which returns the top-l instances with the smallest p-ranks. Clearly, top-(p,l) queries are reverse queries of top-(k,l) queries.■

Interestingly, it is easy to show the following.

Corollary 2.3 (Answers to PT-k and RT-k queries). *Given a set of uncertain objects S, an integer $k > 0$ and a real value $p \in (0,1]$, the answer to a PT-k query with rank parameter k and probability threshold p and that to a RT-k query with rank threshold k and probability threshold p are identical.*

Proof. An instance satisfying the PT-k query must have the p-rank at most k, and thus satisfies the RT-k query. Similarly, an instance satisfying the RT-k query must have the top-k probability at least p, and thus satisfies the PT-k query. ■

PT-k queries and RT-k queries share the same set of parameters: a rank parameter k and a probability threshold. Thus, as shown in Corollary 2.3, when the parameters are the same, the results are identical. For top-(k,l) queries and top-(p,l) queries, even they share the same value on the answer set size constraint l, the answers generally may not be the same since the rank parameter and the probability threshold select different instances.

TID	Rank	Prob.	Top-k probabilities			
			$k=1$	$k=2$	$k=3$	$k=4$
o_1	1	0.5	0.5	0.5	0.5	0.5
o_2	2	0.3	0.15	0.3	0.3	0.3
o_3	3	0.7	0.245	0.595	0.7	0.7
o_4	4	0.9	0.0945	0.45	0.8055	0.9

Table 2.2 Top-k probabilities of a set of tuples.

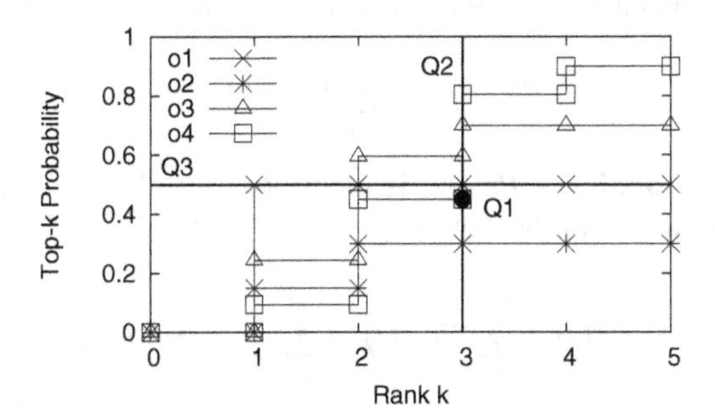

Fig. 2.5 Ranking queries on uncertain tuples.

Example 2.8 (PT-k query and Top-(k,l) query). Consider a set of uncertain instances in Table 2.2. Suppose each instance belongs to one uncertain object and all objects are independent. In Figure 2.5, we plot the top-k probabilities of all instances with respect to different values of k.

A PT-3 query with probability threshold $p = 0.45$ returns instances $\{o_1, o_3, o_4\}$ whose top-3 probabilities are at least 0.45. Interestingly, the PT-3 query with probability threshold $p = 0.45$ can be represented as a point $Q1(3, 0.45)$ in Figure 2.5. As the answers to the query, the top-k probability curves of o_1, o_3 and o_4 lie northeast to $Q1$.

Alternatively, a top-(k,l) query with $k = 3$ and $l = 2$ returns 2 instances $\{o_4, o_3\}$, which have the highest top-3 probabilities. The query can be represented as a vertical line $Q2(k = 3)$ in Figure 2.5. The answer set includes the 2 curves which have the highest intersection points with Q2. ■

Example 2.9 (RT-k query and Top-(p,l) query). Consider the uncertain instances in Table 2.2 again. An RT-3 query with probability threshold $p = 0.45$ returns $\{o_1, o_3, o_4\}$. The answer is the same as the answer to the PT-3 query with the same probability threshold as shown in Example 2.8.

A top-(p,l) query with $p = 0.5$ and $l = 2$ returns $\{o_1, o_3\}$ whose 0.5-ranks are the smallest. The query can be represented as a horizontal line $Q3(probability = 0.5)$ in Figure 2.5. The 2 curves having the leftmost intersections with Q3 are the answers. ■

2.2.3 Ranking Uncertain Objects

At the object level, the rank-k probability and top-k probability are defined as follows.

Definition 2.14 (Object rank-k probability and top-k probability). For an uncertain object O, the **object rank-k probability** $Pr(O,k)$ is the probability that any instance $o \in O$ is ranked at the k-th position in possible worlds according to f, that is

$$Pr(O,k) = \sum_{o \in O} Pr(o,k). \tag{2.3}$$

The **object top-k probability** $Pr^k(O)$ is the probability that any instance in O is ranked top-k in possible worlds, that is

$$Pr^k(O) = \sum_{o \in O} Pr^k(O) = \sum_{o \in O} \sum_{j=1}^{k} Pr(o,j). \tag{2.4}$$

■

The probabilistic ranking queries discussed in Section 2.2.2 can be applied at the object level straightforwardly. Therefore, we skip the definitions of those queries.

2.3 Extended Uncertain Data Models and Ranking Queries

In this section, we develop three extended uncertain data models and ranking queries on those models, to address different application interest.

2.3.1 Uncertain Data Stream Model

As illustrated in Scenario 2 of Example 1.1, the instances of an uncertain object may keep arriving in fast pace and thus can be modeled as a data stream. The instances are generated by an underlying *temporal random variable* whose distribution evolves over time. To keep our discussion simple, we assume a synchronous model. That is, each time instant is a positive integer, and at each time instant t ($t > 0$), an instance is collected for an uncertain data stream. To approximate the current distribution of a temporal random variable, practically we often use the observations of the variable in a recent time window as the sample instances.

Definition 2.15 (Uncertain data stream, sliding window).

An **uncertain data stream** is a (potentially infinite) series of instances $O = o_1, o_2, \cdots$. Given a time instants t ($t > 0$), $O[t]$ is the instance of stream O.

A **sliding window** W_ω^t is a selection operator defined as $W_\omega^t(O) = \{O[i] | (t-\omega) < i \leq t\}$, where $\omega > 0$ is called the **width** of the window.

For a set of uncertain data streams $\mathscr{O} = \{O_1, \cdots, O_n\}$, sliding window $W_\omega^t(\mathscr{O}) = \{W_\omega^t(O_i) | 1 \leq i \leq n\}$. ■

2.3.1.1 Connections with the Uncertain Object Model

The distribution of an uncertain data stream O in a given sliding window W_ω^t is static. Thus, the set of instances $W_\omega^t(O)$ can be considered as an uncertain object. The membership probabilities for instances depend on how the instances are generated from the underlying random variable of $W_\omega^t(O)$. For example, if the instances are drawn using simple random sampling [34], then all instances take the same probability $\frac{1}{\omega}$. On the other hand, using other techniques like particle filtering [35] can generate instances with different membership probabilities. In this book, we assume that the membership probabilities of all instances are identical. Some of our developed methods can also handle the case of different membership probabilities, which will be discussed in Section 6.5.

Definition 2.16 (Uncertain object in a sliding window). Let O be an uncertain data stream. At time instant $t > 0$, the set of instances of O in a sliding window W_ω^t is an uncertain object denoted by $W_\omega^t(O)$ $(1 \leq i \leq n)$, where each instant $o \in W_\omega^t(O)$ has the membership probability $Pr(o) = \frac{1}{\omega}$. ■

In this book, we assume that the distributions of uncertain data streams are independent from each other. Handling correlations among uncertain data streams is an important direction that we plan to investigate as future study that will be discussed in Section 6.5. The uncertain data in a sliding window carries the possible worlds semantics.

Definition 2.17 (Possible worlds of uncertain data streams). Let $\mathscr{O} = \{O_1, \ldots, O_n\}$ be a set of uncertain data streams. A **possible world** $w = \{v_1, \ldots, v_n\}$ in a sliding window W_ω^t is a set of instances such that one instance is taken from the uncertain object of each stream in W_ω^t, i.e., $v_i \in W_\omega^t(O_i)$ $(1 \leq i \leq n)$. The **existence probability** of w is $Pr(w) = \prod_{i=1}^{n} Pr(v_i) = \prod_{i=1}^{n} \frac{1}{\omega} = \omega^{-n}$.

The complete set of possible worlds of sliding window $W_\omega^t(\mathscr{O})$ is denoted by $\mathscr{W}(W_\omega^t(\mathscr{O}))$. ■

Corollary 2.4 (Number of possible worlds). *For a set of uncertain data streams $\mathscr{O} = \{O_1, \ldots, O_n\}$ and a sliding window $W_\omega^t(\mathscr{O})$, the total number of possible worlds is $|\mathscr{W}(W_\omega(t))| = \omega^n$.* ■

When it is clear from the context, we write $\mathscr{W}(W_\omega^t(\mathscr{O}))$ as $\mathscr{W}$ and $W_\omega^t(\mathscr{O})$ as W or W^t for the sake of simplicity.

Time instant	# Time	Speeds at A	Speeds at B	Speeds at C	Speeds at D
$t-2$	00 : 01 : 51	$a_1 = 15$	$b_1 = 6$	$c_1 = 14$	$d_1 = 4$
$t-1$	00 : 16 : 51	$a_2 = 16$	$b_2 = 5$	$c_2 = 8$	$d_2 = 7$
t	00 : 31 : 51	$a_3 = 13$	$b_3 = 1$	$c_3 = 2$	$d_3 = 10$
$t+1$	00 : 46 : 51	$a_4 = 11$	$b_4 = 6$	$c_4 = 9$	$d_4 = 3$
...					

Table 2.3 An uncertain data stream. (Sliding window width $\omega = 3$. W_3^t contains time instant $t-2$, $t-1$ and t. W_3^{t+1} contains time instant $t-1$, t and $t+1$.)

Example 2.10 (Uncertain streams).

As discussed in Example 1.1, speed sensors are deployed to monitor traffic in a road network. The vehicle speed at each monitoring point can be modeled as a ***temporal random variable****. To capture the distribution of such a temporal random variable, a speed sensor at the monitoring point reports the speed readings every* 30 *seconds. Therefore, the speed readings reported by each speed sensor is an* ***uncertain stream****. Each reading is an* ***instance*** *of the stream. A* ***sliding window*** *of length* 3 *at time t contains the last* 3 *readings (that is, the readings in the last* 90 *seconds) of each speed sensor.*

Suppose there are four monitoring points A, B, C and D with speed readings shown in Table 2.3. At time t, sliding window W_3^t contains the records of speeds at time $t-2$, $t-1$ and t. $W_3^t(A) = \{a_1, a_2, a_3\}$ can be modeled as an ***uncertain object****. So are $W_3^t(B)$, $W_3^t(C)$ and $W_3^t(D)$. Each instance in W_3^t takes* ***membership probability*** $\frac{1}{3}$*. There are $3^4 = 81$ possible worlds. Each possible world takes one instance from each object. For example, $\{a_1, b_3, c_2, d_1\}$ is a possible world. The existence probability of each possible world is $(\frac{1}{3})^4 = \frac{1}{81}$.* ■

2.3.1.2 Continuous Probabilistic Threshold Top-*k* Queries

Probabilistic threshold top-*k* queries can be applied on a sliding window of multiple uncertain data streams. We treat the instances of an uncertain data stream falling into the current sliding window as an uncertain object, and rank the streams according to their current sliding window.

Definition 2.18 (Continuous probabilistic threshold top-*k* query). Given a probabilistic threshold top-*k* query Q_p^k, a set of uncertain data streams $\mathscr{O}$, and a sliding window width ω, the **continuous probabilistic threshold top-*k* query** is to, for each time instant t, report the set of uncertain data streams whose top-*k* probabilities in the sliding window $W_\omega^t(\mathscr{O})$ are at least p. ■

Example 2.11 (Continuous Probabilistic Threshold Top-k Queries). Consider the uncertain streams in Table 2.3 with sliding window size $\omega = 3$ and continuous probabilistic threshold top-2 query with threshold $p = 0.5$.

At time instant t, the sliding window contains uncertain objects $W_3^t(A)$, $W_3^t(B)$, $W_3^t(C)$ and $W_3^t(D)$. The top-k probabilities of those uncertain objects are:

$Pr^2(W_3^t(A)) = 1$, $Pr^2(W_3^t(B)) = \frac{2}{27}$, $Pr^2(W_3^t(C)) = \frac{5}{9}$ and $Pr^2(W_3^t(D)) = \frac{10}{27}$. Therefore, the probabilistic threshold top-k query returns $\{A,C\}$ at time instant t.

At time instant $t+1$, the top-k probabilities of the uncertain objects are: $Pr^2(W_3^{t+1}(A)) = 1$, $Pr^2(W_3^{t+1}(B)) = \frac{2}{27}$, $Pr^2(W_3^{t+1}(C)) = \frac{4}{9}$ and $Pr^2(W_3^{t+1}(D)) = \frac{13}{27}$. The probabilistic threshold top-k query returns $\{A\}$ at time instant $t+1$.

The methods of answering probabilistic threshold top-k queries will be discussed in Chapter 6.

2.3.2 Probabilistic Linkage Model

In the basic uncertain object model, we assume that each instance belongs to a unique object, though an object may have multiple instances. It is interesting to ask what if an instance may belong to different objects in different possible worlds. Such a model is useful in probabilistic linkage analysis, as shown in the following example.

Example 2.12 (Probabilistic linkages). Survival-after-hospitalization is an important measure used in public medical service analysis. For example, to obtain the statistics about the death population after hospitalization, Svartbo et al. [36] study survival-after-hospitalization by linking two real data sets, the hospitalization registers and the national causes-of-death registers in some counties in Sweden. Such technique is called record linkage [37], which finds the linkages among data entries referring to the same real-world entities from different data sources. However, in real applications, data is often incomplete or ambiguous. Consequently, record linkages are often uncertain.

Probabilistic record linkages are often used to model the uncertainty. For two records, a state-of-the-art probabilistic record linkage method [37, 38] can estimate the probability that the two records refer to the same real-world entity. To illustrate, consider some synthesized records in the two data sets as shown in Table 2.4. The column probability is calculated by a probability record linkage method.

Two thresholds δ_M and δ_U are often used ($0 \leq \delta_U < \delta_M \leq 1$): when the linkage probability is less than δ_U, the records are considered not-matched; when the linkage probability is between δ_U and δ_M, the records are considered possibly matched; and when the linkage probability is over δ_M, the records are considered matched. Many previous studies focus on building probabilistic record linkages effectively and efficiently.

If a medical doctor wants to know, between John H. Smith *and* Johnson R. Smith, *which patient died at a younger age. The doctor can set the two thresholds $\delta_M = 0.4$ and $\delta_U = 0.35$ and compare the matched pairs of records. Suppose $\delta_M = 0.4$ and $\delta_U = 0.35$, then* John H. Smith *is matched to* J. Smith, *whose age is* 61, *and* Johnson R. Smith *is matched to* J. R. Smith, *whose age is* 45. *Therefore, the medical doctor concludes that* Johnson R. Smith *died at a younger age than* John H. Smith. *Is the answer correct?*

LID	hospitalization registers			causes-of-death registers			Probability
	Id	Name	Disease	Id	Name	Age	
l_1	a_1	John H. Smith	Leukemia	b_1	Johnny Smith	32	0.3
l_2	a_1	John H. Smith	Leukemia	b_2	John Smith	35	0.3
l_3	a_1	John H. Smith	Leukemia	b_3	J. Smith	61	0.4
l_4	a_2	Johnson R. Smith	Lung cancer	b_3	J. Smith	61	0.2
l_5	a_2	Johnson R. Smith	Lung cancer	b_4	J. R. Smith	45	0.8

Table 2.4 Record linkages between the hospitalization registers and the causes-of-death registers.

If we consider all possible worlds corresponding to the set of linkages shown in Table 2.4 (the concept of possible world on probabilistic linkages will be defined in Definition 2.20), then the probability that Johnson R. Smith *is younger than* John H. Smith *is* 0.4*, while that probability that* John H. Smith *is younger than* Johnson R. Smith *is* 0.6*. Clearly, between the two patient,* John H. Smith *died at a younger age than* Johnson R. Smith *with higher probability. How to compute this probability will be discussed in Chapter 7.*

In this example, we can consider each linked pair of records as an uncertain instance and each record as an uncertain object. Two uncertain objects from different data sets may share zero or one instance. Therefore, the uncertain objects may not be independent. We develop the probabilistic linkage model to describe such uncertain data. ■

Let $\mathscr{E}$ be a set of real-world entities. We consider two tables A and B which describe subsets $\mathscr{E}_A, \mathscr{E}_B \subseteq \mathscr{E}$ of entities in $\mathscr{E}$. Each entity is described by at most one tuple in each table. In general, $\mathscr{E}_A$ and $\mathscr{E}_B$ may not be identical. Tables A and B may have different schemas as well.

Definition 2.19 (Probabilistic linkage). Consider two tables A and B, each describing a subset of entities in $\mathscr{E}$, a **linkage function** $\mathscr{L} : A \times B \rightarrow [0,1]$ gives a score $\mathscr{L}(t_A, t_B)$ for a pair of tuples $t_A \in A$ and $t_B \in B$ to measure the likelihood that t_A and t_B describe the same entity in $\mathscr{E}$. A pair of tuples $l = (t_A, t_B)$ is called a **probabilistic record linkage** (or **linkage** for short) if $\mathscr{L}(l) > 0$. $Pr(l) = \mathscr{L}(t_A, t_B)$ is the **membership probability** of l. ■

Given a linkage $l = (t_A, t_B)$, the larger the membership probability $Pr(l)$, the more likely the two tuples t_A and t_B describe the same entity. A tuple $t_A \in A$ may participate in zero, one or multiple linkages. The number of linkages that t_A participates in is called the **degree** of t_A, denoted by $d(t_A)$. Symmetrically, we can define the degree of a tuple $t_B \in B$.

For a tuple $t_A \in A$, let $l_1 = (t_A, t_{B_1}), \cdots, l_{d(t_A)} = (t_A, t_{B_{d(t_A)}})$ be the linkages that t_A participates in. For each tuple $t_A \in A$, we can write a **mutual exclusion rule** $R_{t_A} = l_1 \oplus \cdots \oplus l_{d(t_A)}$ which indicates that at most one linkage can hold based on the assumption that each entity can be described by at most one tuple in each table. $Pr(t_A) = \sum_{i=1}^{d(t_A)} Pr(l_i)$ is the probability that t_A is matched by some tuples in B. Since the linkage function is normalized, $Pr(t_A) \leq 1$. We denote by $R_A = \{R_{t_A} | t_A \in A\}$ the set of mutual exclusion rules for tuples in A. R_{t_B} for $t_B \in B$ and R_B are defined symmetrically.

$(\mathscr{L}, A, B)$ specifies a bipartite graph, where the tuples in A and those in B are two independent sets of nodes, respectively, and the edges are the linkages between the tuples in the two tables.

2.3.2.1 Connections with the Uncertain Object Model

Given a set of probabilistic linkage $\mathscr{L}$ between tuple sets A and B, we can consider each tuple $t_A \in A$ as an uncertain object. For any tuple $t_B \in B$, if there is a linkage $l = (t_A, t_B) \in \mathscr{L}$ such that $Pr(l) > 0$, then t_B can be considered as an instance of object t_A whose membership probability is $Pr(l)$. In contrast to the basic uncertain object model where each instance only belongs to one object, in the probabilistic linkage model, a tuple $t_B \in B$ may be the instance of multiple objects $\{t_{A_1}, \cdots, t_{A_d}\}$, where t_{A_i} is a tuple in A with linkage $(t_{A_i}, t_B) \in \mathscr{L}$ $(1 \leq i \leq d)$. A mutual exclusion rule $R_{t_B} = (t_{A_i}, t_B) \oplus \cdots \oplus (t_{A_d}, t_B)$ specifies that t_B should only belong to one object in a possible world. Alternatively, we can consider each tuple $t_B \in B$ as an uncertain object and a tuple $t_A \in A$ is an instance of t_B if there is a linkage $(t_A, t_B) \in \mathscr{L}$.

A linkage function can be regarded as the summarization of a set of possible worlds.

Definition 2.20 (Possible worlds). For a linkage function $\mathscr{L}$ and tables A and B, let $\mathscr{L}_{A,B}$ be the set of linkages between tuples in A and B. A **possible world** of $\mathscr{L}_{A,B}$, denoted by $W \subseteq \mathscr{L}_{A,B}$, is a set of pairs $l = (t_A, t_B)$ such that (1) for any mutual exclusive rule R_{t_A}, if $Pr(t_A) = 1$, then there exists one pair $(t_A, t_B) \in W$, symmetrically, for any mutual exclusive rule R_{t_B}, if $Pr(t_B) = 1$, then there exists one pair $(t_A, t_B) \in W$; and (2) each tuple $t_A \in A$ participates in at most one pair in W, so does each tuple $t_B \in B$.

$\mathscr{W}_{\mathscr{L},A,B}$ denotes the set of all possible worlds of $\mathscr{L}_{A,B}$. ∎

We study the ranking query answering on probabilistic linkage model in Chapter 7.

2.3.3 *Uncertain Road Network*

As illustrated in Scenario 3 of Example 1.1, the weight of each edge in a graph may be an uncertain object. An uncertain road network is a probabilistic graph defined as follows.

Definition 2.21 (Probabilistic graph). A **probabilistic graph** $G(V, E, W)$ is a simple graph containing a set of vertices V, a set of edges $E \subseteq V \times V$, and a set of weights W defined on edges in E. For each edge $e \in E$, $w_e \in W$ is a real-valued random variable in $(0, +\infty)$, denoting the travel time along edge e. ∎

As discussed in Section 2.1.1, the distribution of w_e is often unavailable and can be estimated by a set of *samples* $\{x_1, \cdots, x_m\}$, where each sample $x_i > 0$ $(1 \leq i \leq m)$ takes a *membership probability* $Pr(x_i) \in (0,1]$ to appear. Moreover, $\sum_{i=1}^{m} Pr(x_i) = 1$.

2.3.3.1 Paths and Weight Distribution

A **simple path** P is a sequence of non-repeated vertices $\langle v_1, \ldots, v_{n+1} \rangle$, where $e_i = (v_i, v_{i+1})$ is an edge in E $(1 \leq i \leq n)$. v_1 and v_{n+1} are called the **start vertex** and the **end vertex** of P, respectively. For the sake of simplicity, we call a simple path a **path** in the rest of the paper. Given two vertices u and v, the complete set of paths between u and v is denoted by $\mathscr{P}_{u,v}$.

For paths $P = \langle v_1, \ldots, v_{n+1} \rangle$ and $P' = \langle v_{i_0}, v_{i_0+1}, \ldots, v_{i_0+k} \rangle$ such that $1 \leq i_0 \leq n+1-k$, P is called a **super path** of P' and P' is called a **subpath** of P. Moreover, $P = \langle P_1, P_2, \ldots, P_m \rangle$ if $P_1 = \langle v_1, \ldots, v_{i_1} \rangle$, $P_2 = \langle v_{i_1+1}, \ldots, v_{i_2} \rangle$, $\ldots$, $P_m = \langle v_{i_{m-1}+1}, \ldots, v_{n+1} \rangle$, $1 < i_1 < i_2 < \cdots < i_{m-1} \leq n$. P_j $(1 \leq j \leq m)$ is called a **segment** of P.

The **weight** of path $P = \langle v_1, \ldots, v_{n+1} \rangle$ is the sum of the weights of all edges in P, that is $w_P = \sum_{i=1}^{n} w_{e_i}$, where w_{e_i} is the weight of edge $e_i = (v_i, v_{i+1})$ with probability mass function $f_{e_i}(x)$. Since each w_{e_i} is a discrete random variable, w_P is also a discrete random variable. A **sample** of P is $x_P = \sum_{i=1}^{n} x_i$, where x_i $(1 \leq i \leq n)$ is a sample of edge $e_i = (v_i, v_{i+1})$. We also write $x_P = \langle x_1, \ldots, x_n \rangle$ where $x_1, \ldots, x_n$ are called the **components** of x_P.

The **probability mass function** of w_P is

$$f_P(x) = Pr[w_P = x] = \sum_{x_1 + \ldots + x_n = x} Pr[w_{e_1} = x_1, \ldots, w_{e_n} = x_n] \tag{2.5}$$

In road networks, the travel time on a road segment e may be affected by the travel time on other roads connecting with e. Therefore, the weights of adjacent edges in E may be correlated. Among all edges in path P, the correlation between the weights w_{e_i} and $w_{e_{i+1}}$ of two adjacent edges e_i and e_{i+1} $(1 \leq i \leq n)$ can be represented using different methods, depending on the types of correlations. To keep our discussion general, in this paper we represent the correlations between w_{e_i} and $w_{e_{i+1}}$ using the joint distribution over the sample pairs $(x_i, x_{i+1}) \in w_{e_i} \times w_{e_{i+1}}$. The **joint probability mass function**[1] of w_{e_i} and $w_{e_{i+1}}$ is $f_{e_i, e_{i+1}}(x_i, x_{i+1}) = f_{e_{i+1}, e_i}(x_{i+1}, x_i) = Pr[w_{e_i} = x_i, w_{e_{i+1}} = x_{i+1}]$. Correspondingly, the **conditional probability** of w_{e_i} given $w_{e_{i+1}}$ is $f_{e_i | e_{i+1}}(x_i | x_{i+1}) = \frac{f_{e_i, e_{i+1}}(x_i, x_{i+1})}{f_{e_{i+1}}(x_{i+1})}$.

Theorem 2.1 (Path weight mass function). *The probability mass function of a simple path $P = \langle v_1, \ldots, v_{n+1} \rangle$ ($e_i = (v_i, v_{i+1})$ for $1 \leq i \leq n$) is*

[1] The joint travel time distribution among connected roads can be obtained from roadside sensors. The sensors report the speeds of vehicles passing the sensors. The speeds can be transformed into travel time. A set of travel time values reported by sensors at the same time is a sample of the joint travel time distribution.

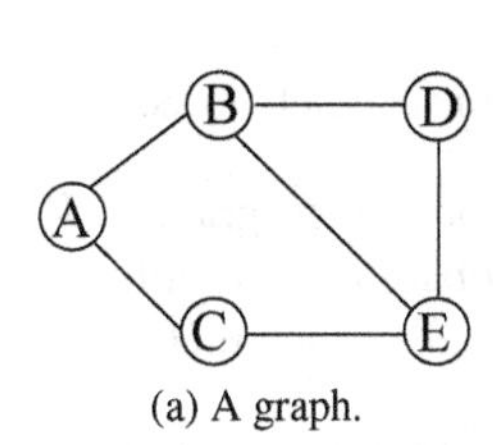

(a) A graph.

Edge e_i	Weight: value(probability) x_{i1}	x_{i2}	x_{i3}
e_1: AB	10(0.3)	15(0.3)	20(0.4)
e_2: AC	5(0.2)	10(0.3)	15(0.5)
e_3: BD	20(0.4)	25(0.4)	30(0.2)
e_4: BE	5(0.2)	25(0.6)	40(0.2)
e_5: CE	10(0.5)	20(0.1)	45(0.4)
e_6: DE	10(0.3)	20(0.6)	50(0.1)

(b) Probabilistic weights of edges.

	20	25	30
10	0.15	0.15	0
15	0.15	0.15	0
20	0.1	0.1	0.2

(c) $f_{e_1,e_3}(x_{1i},x_{3j})$.

	10	20	50
20	0.1	0.2	0.1
25	0.1	0.3	0
30	0.1	0.1	0

(d) $f_{e_3,e_6}(x_{3i},x_{6j})$.

Fig. 2.6 A probabilistic graph.

$$f_P(x) = \sum_{x_1+\ldots+x_n=x} \frac{\prod_{i=1}^{n-1} f_{e_i,e_{i+1}}(x_i,x_{i+1})}{\prod_{j=2}^{n-1} f_{e_j}(x_j)} \tag{2.6}$$

Proof. Since P is a simple path, each edge $e_i \in P$ $(1 \leq i \leq)$ is only adjacent with e_{i-1} (if $i > 1$) and e_{i+1} (if $i < n$) in P. Therefore, given w_{e_i}, the weights $w_{e_1},\ldots,w_{e_{i-1}}$ are conditionally independent on $w_{e_{i+1}},\ldots,w_{e_n}$. Equation 2.6 follows with basic probability theory. ■

In sequel, the **cumulative distribution function** of w_P is

$$F_P(x) = Pr[w_P \leq x] = \sum_{0<x_i\leq x} f_P(x_i) \tag{2.7}$$

We call $F_P(x)$ the **x-weight probability** of path P.

Example 2.13 (Probabilistic graph and paths). A probabilistic graph is shown in Figure 2.6, where the weight of each edge is represented by a set of samples and their membership probabilities.

Path $P = \langle A,B,D,E\rangle$ consists of edges AB, BD and DE. The joint probabilities of (w_{AB},w_{BD}) and (w_{BD},w_{DE}) are shown in Figures 2.6(c) and 2.6(d), respectively. The probability that $w_P = 45$ is

$$\begin{aligned}
& Pr[w_P = 45] \\
&= Pr[w_{e_1} = 15, w_{e_3} = 20, w_{e_6} = 10] + Pr[w_{e_1} = 10, w_{e_3} = 25, w_{e_6} = 10] \\
&= \frac{f_{e_1,e_3}(15,20)\times f_{e_3,e_6}(20,10)}{f_{e_3}(20)} + \frac{f_{e_1,e_3}(10,25)\times f_{e_3,e_6}(25,10)}{f_{e_3}(25)} \\
&= 0.075
\end{aligned}$$

■

2.3.3.2 Path Queries

We formulate the probabilistic path queries on uncertain road networks.

Definition 2.22 (Probabilistic path queries). Given probabilistic graph $G(V,E,W)$, two vertices $u,v \in V$, a weight threshold $l > 0$, and a probability threshold $\tau \in (0,1]$, a **probabilistic path query** $Q_l^\tau(u,v)$ finds all paths $P \in \mathscr{P}_{u,v}$ such that $F_P(l) \geq \tau$. ■

There can be many paths between two vertices in a large graph. Often, a user is interested in only the "best" paths and wants a ranked list. Thus, we define weight- and probability-threshold top-k path queries.

Definition 2.23 (Top-k probabilistic path queries). Given probabilistic graph $G(V,E,W)$, two vertices $u,v \in V$, an integer $k > 0$, and a weight threshold $l > 0$, a **weight-threshold top-k path query** $WTQ_l^k(u,v)$ finds the k paths $P \in \mathscr{P}_{u,v}$ with the largest $F_P(l)$ values.

For a path P, given **probability threshold** $\tau \in (0,1]$, we can find the smallest weight x such that $F_P(x) \geq \tau$, which is called the **τ-confident weight**, denoted by

$$F_P^{-1}(\tau) = \min\{x | x \in w_P \wedge Pr[w_P \leq x] \geq \tau\} \tag{2.8}$$

A **probability-threshold top-k path query** $PTQ_\tau^k(u,v)$ finds the k paths $P \in \mathscr{P}_{u,v}$ with the smallest $F_P^{-1}(\tau)$ values. ■

Example 2.14 (Path Queries). In the probabilistic graph in Figure 2.6, there are 4 paths between A and D, namely $P_1 = \langle A,B,D \rangle$, $P_2 = \langle A,B,E,D \rangle$, $P_3 = \langle A,C,E,B,D \rangle$, and $P_4 = \langle A,C,E,D \rangle$. Suppose the weights of all edges are independent in this example.

Given a weight threshold $l = 48$ and a probability threshold $\tau = 0.8$, a probabilistic path query Q_l^τ finds the paths whose weights are at most 48 of probability at least 0.8. According to the cumulative distribution functions of the paths, we have $F_{P_1}(48) = 0.92$, $F_{P_2}(48) = 0.14$, $F_{P_3}(48) = 0.028$, and $F_{P_4}(48) = 0.492$. Thus, the answer is $\{P_1\}$.

The weight-threshold top-3 path query $WTQ_l^3(A,D)$ finds the top-3 paths P having the largest 48-weight probability values $F_P(48)$. The answer to $WTQ_l^2(A,D)$ is $\{P_1,P_4,P_2\}$.

The probability-threshold top-3 path query $PTQ_\tau^3(A,D)$ finds the top-3 paths P having the smallest 0.8-confidence weights $F_P^{-1}(0.8)$. Since $F_{P_1}(40) = 0.7$ and $F_{P_1}(45) = 0.92$, the smallest weight that satisfies $F_P(x) \geq 0.8$ is 45. Thus, $F_{P_1}^{-1}(0.8) = 45$. Similarly, we have $F_{P_2}^{-1}(0.8) = 75$, $F_{P_3}^{-1}(0.8) = 105$, and $F_{P_4}^{-1}(0.8) = 75$. Therefore, the answer to $PTQ_\tau^2(A,D)$ is $\{P_1,P_2,P_4\}$. ■

To keep our presentation simple, in the rest of the book, we call probabilistic path queries, weight- and probability-threshold top-k queries as **path queries**, **WT top-k queries**, and **PT top-k queries**, respectively.

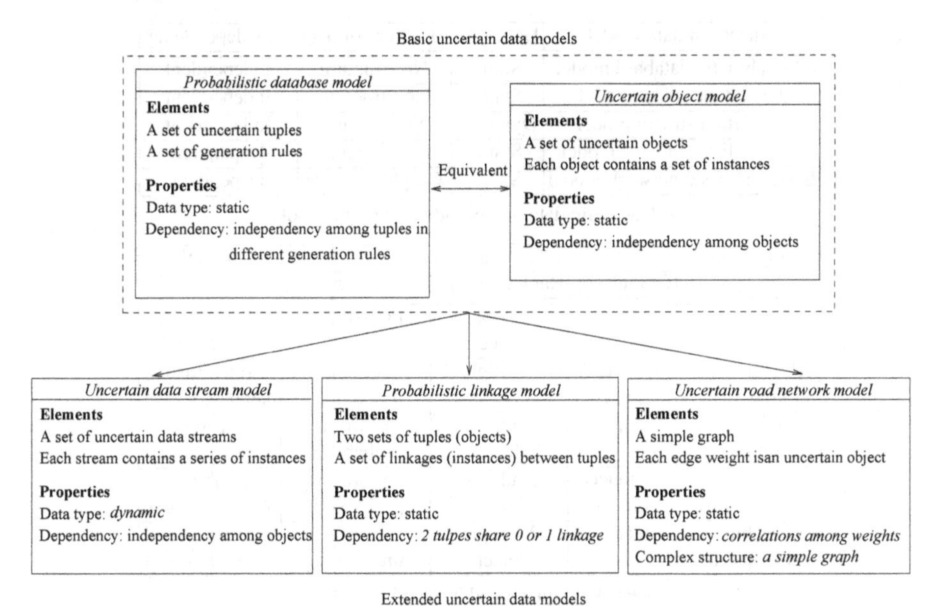

Fig. 2.7 The basic and extended uncertain data models adopted in this book.

2.3.3.3 Connections with the Uncertain Object Model

In the uncertain road network model, the weight of each edge can be considered as an uncertain object with a set of instances. The weight of a path is an aggregate sum of the uncertain objects corresponding to the edges in the path. Therefore, a probabilistic query essentially ranks a set of aggregate sums of uncertain objects.

2.4 Summary

In this chapter, we reviewed two basic uncertain data models, the uncertain object model and the probabilistic database model, as well as the possible worlds semantics that have been extensively adopted in other research on uncertain data.

Moreover, we proposed three extended uncertain data models that extend the uncertain object model from different aspects.

- The *uncertain data stream model* captures the dynamic nature of uncertain data. Each uncertain data stream is a series of (potentially) unlimited instances. Given a sliding window that selects the time span of interest, the instances of each uncertain data stream in the sliding window can be considered as an uncertain object. The uncertain data stream model suits the needs of applications that in-

Uncertain data model	Data type	Structure of data	Data dependency
Probabilistic database model	Static	No structure	Independent
Uncertain object model	Static	No Structure	Independent
Uncertain stream model	Dynamic	No structure	Independent
Probabilistic linkage model	Static	Tree structure	Dependent
Uncertain road network model	Static	Graph structure	Dependent

(a) Uncertain data models adopted in this book.

Problem	Ranking query types			Model
	Granularity	Ranking scope	Ranking criteria	
Typicality queries	Instance	Single object	Probability	Uncertain data model
Probabilistic ranking queries	Instance/ object	Multiple objects	Score & probability	Probabilistic database model
Top-k stream monitoring	Object	Multiple objects	Score & probability	Uncertain stream model
Linkage ranking queries	Instance/ object	Multiple objects	Score & probability	Probabilistic linkage model
Probabilistic path queries	Object set	Multiple objects	Score & probability	Uncertain road network model

(b) Ranking queries addressed in this book.

Table 2.5 Uncertain data models and ranking queries discussed in this book.

volve uncertain data with evolving distributions, such as traffic monitoring and environmental surveillance.

- The *probabilistic linkage model* introduces dependencies among different uncertain objects. It contains two object sets $\mathcal{O}_A$ and $\mathcal{O}_B$ and a set of linkages $\mathcal{L}$. Each linkage matches one object in $\mathcal{O}_A$ with another object in $\mathcal{O}_B$ with a confidence value indicating the how likely the two objects refer to the same real-life entity. Two objects from different object sets may share one instance. The probabilistic linkage model finds important applications in data integration.
- The *uncertain road network model* considers a set of uncertain objects in a simple graph. The weight of each edge in a simple graph is an uncertain object represented by a set of instances. The weights of adjacent edges may involve dependencies. A probabilistic path query finds the optimal paths between two end vertices that have small weights with high confidence. The uncertain road network model is important in applications like real-time trip planning.

Figure 2.7 and Table 2.5(a) summarize the five uncertain data models and their relationship.

Last, we formulated five ranking problems on uncertain data (listed in Table 2.5(b)) and discussed the semantics as well as the potential challenges in query evaluation. Chapters 4 to 8 will discuss the query answering techniques for the proposed ranking problems.

Chapter 3
Related Work

In this chapter, we review the existing studies related to ranking queries on uncertain data. First, we introduce the state-of-the-art studies on uncertain data modeling and processing as well as database ranking queries. Then, we discuss the related work on each of the proposed problems in this book.

3.1 Uncertain Data Processing

Modeling and querying uncertain data has been a fast growing research direction [39, 13, 40, 7] and receives increasing attention. In this section, we review the studies realted to uncertain data modeling and query processing.

3.1.1 Uncertain Data Models and Systems

Various models for uncertain and probabilistic data have been developed in literature. The working model for uncertain data proposed in [7] describes the existence probability of a tuple in an uncertain data set and the constraints (i.e., exclusiveness) on the uncertain tuples. One extensively used model described in [7] is the probabilistic database model discussed in Section 2.1.2. A probabilistic database comprises of multiple probabilistic tables. A probabilistic table contains a set of tuples, where each tuple is associated with a membership probability. A probabilistic table may also come with some generation rules to capture the dependencies among tuples, where a generation rule specifies a set of exclusive tuples, and each tuple is involved in at most one generation rule.

Another popularly used model is the uncertain object model [18, 19, 20, 21] discussed in Section 2.1. An uncertain object is conceptually described by a probability density function in the data space. In a scenario where the probability density function of an uncertain object is unavailable, a set of samples (called instances) are collected to approximate the probability distribution.

In [41, 42, 43, 44], probabilistic graphical models are used to represent correlations among probabilistic tuples. Moreover, Sen *et al.* [45, 46] studied the problem of compact representation of correlations in probabilistic databases by exploiting the shared correlation structures. Probabilistic graphical models are reviewed in Section 3.6.2.

In addition, uncertainty in data integration is studied in [9, 2], where probabilistic schema mapping is modeled as a distribution over a set of possible mappings between two schemas.

On the system level, Orion [47, 48] is an uncertain database management system that supports the attribute and tuple level uncertainty with arbitrary correlations. Three basic operations are defined to perform selection and to compute marginal distributions and joint distributions. Other relational operations can be defined based on the three basic operations. Both continuous and discrete uncertainty is handled in Orion [47, 48].

3.1.2 Probabilistic Queries on Uncertain Data

Cheng *et al.* [18] provided a general classification of probabilistic queries and evaluation algorithms over uncertain data sets. Different from the query answering in traditional data sets, a probabilistic quality estimate was proposed to evaluate the quality of results in probabilistic query answering. Dalvi and Suciu [11] proposed an efficient algorithm to evaluate arbitrary SQL queries on probabilistic databases and rank the results by their probability. Later, they showed in [49] that the complexity of evaluating conjunctive queries on a probabilistic database is either PTIME or #*P*-complete.

Chen *et al.* [50] studied aggregate queries on data whose attributes may take "partial values", where a "partial value" is a set of possible values with only one being true. The answer to an aggregate query involving such attributes is also in the form of a "partial value". In [51], the efficient evaluation of `sum`, `count`, `min` and `max` queries on probabilistic data is studied based on Monte Carlo simulation. Only the top-k answers to the query with the highest probabilities are returned. Burdick *et al.* [52, 53, 54] extended the OLAP model on imprecise data. Different semantics of aggregation queries are considered over such data. Following with the possible worlds semantics model [23, 24, 7, 12], the answer to an aggregation query is represented as an answer random variable with certain probability distribution over a set of possible values. In [55], several one pass streaming algorithms are proposed to estimate statistical aggregates of a probabilistic data stream, which contains a (potentially infinite) sequence of uncertain objects.

Cheng *et al.* [56] explored the join queries over data sets with attribute level uncertainty, where the values of a tuple in the join attributes are probability distributions in a set of value intervals. In [57], join queries are studied on data sets with tuple level uncertainty, where each tuple in a table is associated with a membership probability. Kimelfeld and Sagiv [58] studied the maximal join queries on probabilistic data, where only the answers whose probabilities are greater than a threshold and are not contained by any other output answers are returned.

3.1.3 Indexing Uncertain Data

There are two categories of indexes on uncertain data. The first category is for uncertain numeric data, such as sensor values or locations of moving objects. Tao *et al.* [20, 59] proposed a U-tree index to facilitate probabilistic range queries on uncertain objects represented by multi-dimensional probability density functions. Ljosa *et al.* [60] developed an APLA-tree index for uncertain objects with arbitrary probability distributions. An APLA for each object is an adaptive piecewise linear approximation and can be regarded as a time series. All those time series are organized by an APLA-tree in a hierarchical manner. Besides, Cheng *et al.* [19] developed PTI, probability thresholding index, to index uncertain objects with one dimensional uncertain values. Bohm *et al.* [61] developed an index for uncertain objects whose probability density functions are Gaussian functions $N(\mu,\sigma)$, by treating each object as a two dimensional points (μ,σ) and indexing the points in an R-tree.

The second category is for uncertain categorical data, such as RFID data or text labels. For example, Singh *et al.* [62] extended the inverted index and signature tree to index uncertain categorical data.

3.2 Ranking (Top-k) Queries

There are numerous existing methods for answering top-k queries on static data. The threshold algorithm (TA) [6] is one of the fundamental algorithms. TA first sorts the values in each attribute and then scans the sorted lists in parallel. A "stopping value" is maintained, which acts as a threshold to prune the tuples in the rest of the lists if they cannot have better scores than the threshold. Several variants of TA have been proposed, such as [4]. [29] provides a comprehensive survey about the ranking queries and evaluation algorithms. The recent developments and extensions of top-k query answering include using views to answer top-k queries efficiently [3], removing redundancy in top-k patterns [63], applying multidimensional analysis in top-k queries [64], continuous monitoring of top-k queries over a sliding window [5], and so forth.

3.2.1 Distributed Top-k Query Processing

Distributed top-k query processing focuses on reducing communication cost while providing high quality answers. [65] studies top-k monitoring queries which continuously report the k largest values from data streams produced at physically distributed locations. In [65], there are multiple logical data objects and each object is associated with an overall logical data value. Updates to overall logical data values arrive incrementally over time from distributed locations. Efficient techniques are proposed to compute and maintain the top-k logical data objects over time with low communication cost among distributed locations and a bounded error tolerance. In [66], an algorithmic framework is proposed to process distributed top-k queries, where the index lists of attribute values are distributed across a number of data peers. The framework provides high quality approximation answers and reduces network communication cost, local peer load, and query response time.

3.3 Top-k Typicality Queries

Aside from the studies reviewed in Sections 3.1 and 3.2, the top-k typicality queries that will be discussed in Chapter 4 is also related to the previous work in the following aspects: typicality in psychology and cognitive science, the k-median problem, typicality probability and spatially-decaying aggregation.

3.3.1 Typicality in Psychology and Cognitive Science

Typicality of objects has been widely discussed in psychology and cognitive science [67, 68]. People judge some objects to be "better examples" of a concept than others. This is known as the *graded structure* [69] of a category. Generally, the graded structure is a continuum of category representativeness, beginning with the most typical members of a category and continuing through less typical members to its atypical members.

There are several determinants of graded structure. One determinant is the *central tendency* [70] of a category. Central tendency is either one or several very representative exemplar(s), either existing in the category or not. An exemplar's similarities to the central tendency determine its typicality in this category. Another determinant of typicality is the *stimulus similarity* [71]. Generally, the more similar an instance is to the other members of its category, and the less similar it is to members of the contrast categories, the higher the typicality rating it has.

The prototype view [72] suggests that a concept be represented by a prototype, such that objects "closer to" or "more similar to" the prototype are considered to be better examples of the associated concept. The exemplar view [73] is an alternative to the prototype view that proposes using real objects as exemplars instead of abstract prototypes that might not exist in real life. Finally, the schema view [74] improves the prototype view by modeling concepts in schema theory and artificial intelligence knowledge representation.

Feature-frequency model defines typicality from a different scope [75]. A typical member of a category will share many attributes with other members of the category and few attributes with members of other categories. An attribute can be scored based on how many members possess that attribute. A family resemblance score for each member sums up the numerical scores of all attributes possessed by that member. A category member with a higher family resemblance score is considered more typical.

Although typicality has not been used before in query answering on large databases, the idea of typicality was recently introduced into ontology design and conceptual modeling [76], which are generally related to database design.

3.3.1.1 How Is Our Study Related?

Our typicality measures are in the general spirit of typicality measures used in psychology and cognitive science. As suggested by the previous studies in psychology and cognitive science, typicality measures may vary in different applications. In our study, we propose simple typicality, discriminative typicality, and representative typicality for different application requirements.

Studies on typicality in psychology and cognitive science often do not address the concerns about efficient query answering from large databases. Complementary to those studies, we focus on efficient query answering.

3.3.2 The (Discrete) k-Median Problem

Finding typical objects is broadly related to the k-median problem in computational geometry. Given a set S of n points, the *k-median problem* is to find a set M of k points minimizing the sum of distances from all points in S to M. Points in M are called the medians of S. Under the constraint that points in M belong to S, it is known as the *discrete k-median problem*. When $k = 1$, we can find the exact median in $O(n^2)$ time. When k is an arbitrary input parameter, the discrete k median problem on any distance metric is NP-hard [28].

Several approximation algorithms have been proposed to compute the approximate 1-median efficiently. [77] proposes a quad-tree based data structure to support finding the approximate median with a constant approximation ratio in $O(n \log n)$ time. A randomized algorithm is proposed in [78], which computes the approximate median in linear time. Although the approximation ratio cannot be bounded, it performs well in practice. [79] provides a $(1+\delta)$-approximation algorithm with runtime $O(n/\delta^5)$ based on sufficiently large sampling. [80] proposes an algorithm to solve the median problem in L_1 metric in $O(n \log n)$ time.

3.3.2.1 How Is Our Study Related?

The top-k simple typicality query and the discrete k-median problem both want to find the instances in a set of instances optimizing the scores with respect to their relationship to other instances. However, as will be clear in Chapter 4, the functions to optimize are different. The methods of the discrete k-median problem cannot be applied directly to answer top-k typicality queries.

Moreover, in discrete k-median problem, there is no ranking among the k median objects. The top-k representative typicality queries as defined will return k objects in an order.

3.3.3 Clustering Analysis

Clustering analysis partitions a set of data objects into smaller sets of similar objects. [81] is a nice survey of various clustering methods.

The clustering methods can be divided into the following categories. The *partitioning methods* partition the objects into k clusters and optimize some selected partitioning criterion, where k is a user specified parameter. K-means [82], K-medoids [83] and CLARANS [84] are examples of this category. The *hierarchical methods* perform a series of partitions and group data objects into a tree of clusters. BIRCH [85], CURE [86] and Chameleon [87] are examples of hierarchical methods. The *density-based methods* use a local cluster criterion and find the regions in the data space that are dense and separated from other data objects by regions with lower density as clusters. The examples of density-based methods include DBSCAN [88], OPTICS [89] and DENCLUE [90]. The *grid-based methods* use multi-resolution grid data structures and form clusters by finding dense grid cells. STING [91] and CLIQUE [92] are examples of grid-based methods.

3.3.3.1 How Is Our Study Related?

Typicality analysis and clustering analysis both consider similarity among objects. However, the two problems have different objectives. Clustering analysis focuses on partitioning data objects, while typicality analysis aims to find representative instances.

In some studies, cluster centroids are used to represent the whole clusters. However, in general the centroid of a cluster may not be a representative point. For example, medians are often considered as cluster centroids in partitioning clustering methods, but they are not the most typical objects as shown in Chapter 4.

In the density-based clustering method DBSCAN [88], the concept of "core point" is used to represent the point with high density. For a core point o, there are at least $MinPts$ points lying within a radius Eps from o, where $MinPts$ and Eps

are user input parameters. However, "core points" are significantly different from "typical points" in the following two aspects.

First, a "core point" may not be typical. Consider an extreme case where there are two groups of points: the first group of points lie close to each other with a size much larger than $MinPts$, while the second group only contain $MinPts$ points lying within a radius Eps from a point o that are far away from the points in the first group. then, o is a core point but it is not typical at all. Second, a typical point may not be a "core point", either. It is possible that a typical point does not have $MinPts$ points lying within a distance Eps from it, but it still has a high typicality score. A comparison between clustering and typicality analysis on real data sets is given in Chapter 4.

It is possible to extend the existing clustering methods to answer typicality queries, by defining the most typical object in a cluster as the centroid and using the maximal group typicality of clusters as the clustering criteria, which is in the same spirit as our typicality query evaluation algorithms.

3.3.4 Other Related Models

Typicality probability [93, 94] in statistical discriminant analysis is defined as the Mahalanobis distance between an object and the centroid of a specified group, which provides an absolute measure of the degree of membership to the specified group.

Spatially-decaying aggregation [95, 96] is defined as the aggregation values influenced by the distance between data items. Generally, the contribution of a data item to the aggregation value at certain location decays as its distance to that location increases. Nearly linear time algorithms are proposed to compute the ε-approximate aggregation values when the metric space is defined on a graph or on the Euclidean plane.

3.3.4.1 How Is Our Study Related?

Discriminant analysis mainly focuses on how to correctly classify the objects. It does not consider the typicality of group members. Our definition of discriminative typicality combines both the discriminability and the typicality of the group members, which is more powerful in capturing the "important" instances in multi-class data sets. Moreover, [93, 94] do not discuss how to answer those queries efficiently on large data sets.

Spatially-decaying sum with *exponential decay function* [95, 96] is similar to our definition of simple typicality. However, in [95, 96], the spatially-decaying aggregation problem is defined on graphs or Euclidean planes, while we assume only a generic metric space. The efficiency in [95, 96] may not be carried forward to the more general metric space. The techniques developed developed in Chapter 4 may be useful to compute spatially-decaying aggregation on a general metric space.

When typicality queries are computed on graphs or Euclidean planes, some ideas in [95, 96] may be borrowed.

3.4 Probabilistic Ranking Queries

In this section, we first review the recent proposals of ranking queries and evaluation algorithms on uncertain data. Then, we link the problem of ranking uncertain data to the counting principle in probability and statistics, which provides more insights into the ranking uncertain data problem.

3.4.1 Top-k Queries on Uncertain Data

Generally, ranking queries on uncertain data can be classified into the following two categories.

3.4.1.1 Category I: Extensions of traditional ranking queries

The first category is extensions to traditional ranking queries on certain data. That is, given a traditional ranking query with an objective function, all tuples are ranked based on the objective function. Since the results to the query may be different in different possible worlds, various queries capture and summarize the results using different criteria.

Soliman *et al.* [17] proposed two types of ranking queries: U-Top$k$ queries and U-KRanks queries. The answer to a U-Top$k$ query is always a top-k tuple list in some valid possible worlds, and the exact positions of the tuples in the list are preserved. A U-KRanks query finds the tuple of the highest probability at each ranking position. The tuples in the results of a U-KRanks query may not form a valid top-k tuple list in a possible world, though a U-KRanks query always returns k tuples. A tuple may appear more than once in the answer set if it has the highest probability values to appear in multiple ranking positions, respectively. Lian and Chen developed the spatial and probabilistic pruning techniques for U-KRanks queries [97]. Simultaneously with our study, Yi *et al.* [98] proposed efficient algorithms to answer U-Top$k$ queries and U-KRanks queries. Silberstein *et al.* [32] model each sensor in a sensor network as an uncertain object whose values follow some unknown distribution. Then, a top-k query in the sensor network returns the top-k sensors such that the probability of each sensor whose values are ranked top-k in any timestep is the greatest. A sampling-based method collects all values in the network as a sample at randomly chosen timesteps, and the answer to a top-k query is estimated using the samples.

More recently, Cormode *et al.* [99] proposed to rank probabilistic tuples by expected ranks. The expected rank of a tuple t is the expectation of t's ranks in all possible worlds. The rank of t in a possible world W is defined as the number of tuples ranked higher than t in W. If t does not appear in W, then its rank in W is defined as $|W|$. Ranking by expected ranks cannot capture the semantics of the probabilistic ranking queries discussed in this book. For example, consider a probabilistic table containing three tuples t_1, t_2 and t_3, with membership probabilities 0.6, 1 and 1, respectively. Suppose the ranking order on the three tuples based on their scores is $t_1 \prec t_2 \prec t_3$. There are two possible worlds $W_1 = \{t_1, t_2, t_3\}$ and $W_2 = \{t_2, t_3\}$, with probabilities 0.6 and 0.4, respectively. The expected rank of t_1 is $0 \times 0.6 + 2 \times 0.4 = 0.8$. The expected ranks of t_2 and t_3 can be computed similarly, which are 0.6 and 1.6, respectively. A top-1 query based on expected ranks returns t_2 as the result, since t_2 has the smallest expected rank. However, is t_2 the most likely tuple to be ranked top-1? The top-1 probabilities of t_1, t_2 and t_3 are 0.6, 0.4 and 0, respectively. Clearly, t_1 has the highest probability to be ranked top-1. A top-(k,l) query with $k = 1$ and $l = 1$ returns t_1 as the result.

Li *et al.* [100] discussed the problem of ranking distributed probabilistic data. The goal is to minimize the communication cost while retrieve the top-k tuples with expected ranks from distributed probabilistic data sets. In [101], ranking in probabilistic databases is modeled as a multi-criteria optimization problem. A general ranking function of a tuple t is defined as the weighted sum of the position probabilities of t. This allows users to explore the possible ranking functions space. Moreover, how to learn the weight parameters of different position probabilities from user preference data was discussed.

3.4.1.2 Category II: Extensions of general traditional queries

The second category is to use probability to rank answers to a query on uncertain data. That is, given a query on uncertain data, results to the query are ranked according to their probabilities. Such probability is called output probabilities. In [51], Ré *et al.* considered arbitrary SQL queries and the ranking is on the probability that a tuple satisfies the query instead of using a ranking function. [51] and our study address essentially different queries and applications. Meanwhile, Zhang and Chomicki developed the global top-k semantics on uncertain data which returns k tuples having the largest probability in the top-k list, and gave a dynamic programming algorithm [33].

3.4.1.3 How Is Our Study Related?

In this book, we study a class of ranking queries belong to Category I, probabilistic ranking queries (including PT-k queries and top-(k,l) queries) and reverse probabilistic ranking queries (including RT-k queries and top-(r,l) queries). Those queries have been defined in Chapter 2. Among them, reverse probabilistic ranking queries have not been considered by any previous studies.

We proposed [30, 31] probability threshold top-k queries that find the tuples whose probabilities of being ranked top-k are at least p, where p is a user specified probability threshold. Probabilistic threshold top-k queries bear different semantics from U-Top$k$ queries and U-KRanks queries. Consider the following example. If there are three sensors A, B and C deployed at different locations. At time $9AM$, three records r_A, r_B, and r_C are reported from those sensors with associated confidences: $r_A = 110km/h$ with $Pr(r_A) = 0.1$, $r_B = 100km/h$ with $Pr(r_B) = 0.4$, and $r_C = 90km/h$ with $Pr(r_C) = 0.8$. What are the top-2 speeding locations?

- A U-Top2 query reports C as the answer, since $\langle r_C \rangle$ is the most probably top-2 list in all possible worlds, whose probability is 0.432.
- A U-2Ranks query reports C as the most probably 1-st speeding location with confidence 0.432. For the 2-nd speeding location, C is reported again with confidence 0.288.
- A probabilistic threshold top-2 query with probability threshold $p = 0.3$ returns B and C as the 2 locations whose top-2 probabilities are no smaller than p. Their top-2 probabilities are $Pr^2(r_B) = 0.4$ and $Pr^2(r_C) = 0.72$.

Therefore, location B has a probability of 0.4 of being ranked in the top-2 speeding locations. But it cannot be reported by U-Top$k$ queries or U-kRanks queries. In the speed monitoring application, users are more interested in the individual locations with a high probability of being ranked top-k. The co-occurrence of speeding locations in the top-k list or the speeding location at certain ranking position may not be important. Therefore, probabilistic threshold top-k queries are more appropriate than U-Top$k$ queries and U-kRanks queries in this application scenario.

Moreover, we develop efficient query answering algorithms and effective index structures for the proposed queries. Our unique prefix sharing technique and three pruning techniques can greatly improve the efficiency in query answering. It is worth noting that our algorithm can be used to answer U-KRanks query straightforwardly, while their algorithm may not be used to handle PT-k query directly.

3.4.2 Poisson Approximation

The problem of answering probability ranking queries is also related to Poisson trials and the Poisson binomial distribution in probability and statistics.

A Poisson trial is an experiment whose outcome is randomly selected from two possible outcomes, "success" or "failure". Let $X_1, \ldots, X_n$ be n independent Bernoulli random trials. For each trial X_i, the success probability is p_i $(1 \leq i \leq n)$. The experiment is called Bernoulli trials if the trials are identical and thus the success probability of all the trials are the same. Otherwise, the experiment is called Poisson trials [102].

The sum $X = \sum_{i=1}^{n} X_i$ is the total number of successes in n independent Bernoulli trials. X follows a binomial distribution for identical Bernoulli trials, and a Poisson-binomial distribution for Poisson trials [103]. The exact distribution of $Pr(X = i)$ can be calculated recursively using the Poisson binomial recurrence [103].

Given an uncertain table, a generation rule can be viewed as a Bernoulli trial, if we consider the appearance of a tuple as a "success". The probability of a rule is the success probability of the corresponding trial. The probability of a tuple t to be ranked top-k is the probability that t appears and there are fewer than k successes appear before t.

3.4.2.1 How Is Our Study Related?

Some results of Poisson trials can be used in answering probability threshold top-k queries. However, the study of Poisson trials in probability theory does not address the concerns on efficient query answering for large databases. Moreover, multi-tuple generation rules pose new challenges.

In our study, we develop several techniques to process generation rules efficiently. Pruning rules are also proposed to achieve early stop without scanning the whole table, which significantly improves the efficiency in query answering.

3.5 Uncertain Streams

In this section, we review the existing work highly related to ranking query monitoring on uncertain data streams and point out the differences.

3.5.1 Continuous Queries on Probabilistic Streams

To the best of our knowledge, [104, 105, 106, 43] are the only existing studies on continuous queries on probabilistic data streams, which are highly related to our study.

In [104], a probabilistic data stream is a (potentially infinite) sequence of uncertain tuples, where each tuple is associated with a membership probability p $(0 < p \leq 1)$, meaning that the tuple takes a probability p to appear in an instance (i.e., a possible world) of the probabilistic stream. It is assumed that tuples are independent from each other. Conventional stream sketching methods are extended to such probabilistic data streams to approximate answers to complex aggregate queries.

Jayram *et al.* [105, 106] adopted a different probabilistic data stream model. A probabilistic data stream contains a (potentially infinite) sequence of uncertain objects, where each uncertain object is represented by a set of instances and each instance carries a membership probability. An uncertain object arrives in whole at a time and does not change after the arrival. In other words, uncertain objects do not evolve over time. New uncertain objects keep arriving. Several one pass streaming algorithms are proposed to estimate the statistical aggregates of the probabilistic data.

Most recently, Kanagal and Deshpande [43] proposed a probabilistic sequence model that considers the temporal and spatial correlations among data. Given a set of uncertain attributes $(A_1, \cdots, A_m)$, each uncertain attribute A_i $(1 \leq i \leq m)$ is a discrete random variable in domain $dom(A_i)$ whose distribution is evolving over time. A probabilistic sequence contains, for each time instant t, an instance $(v_1^t, \cdots, v_m^t)$ for $(A_1, \cdots, A_m)$, where each v_i^t $(1 \leq i \leq m)$ is a random variable in domain $dom(A_i)$ with certain probability distribution. It is a Markov sequence since the random variables at t only depends on the random variables at $t-1$. Graphical models are used to describe the correlations among the random variables in two consecutive instants. Query answering is considered as inferences over the graphical models.

3.5.1.1 How Is Our Study Related?

Our study is different from [104, 105, 106, 43] in following two important aspects. First, the uncertain stream model proposed in this book is the substantially different from the ones proposed before. In the probabilistic sequence model proposed in [43], each element in the stream is a random variable (distribution). While we model an uncertain stream as a series of sample instances generated by a temporal random variable. The set of random variables (i.e., uncertain objects) are fixed. The distributions of those random variables evolve over time. Our model handles some application scenarios that are not covered by the models in [104, 105, 106, 43].

Second, we focus on continuous probabilistic threshold top-k queries on sliding windows, a novel type of queries on uncertain data streams that have not been addressed before. [104, 105, 106] deal with aggregates on a whole stream. The operators discussed in [43] cannot be directly used to answer continuous probabilistic threshold top-k queries.

3.5.2 Continuous Ranking and Quantile Queries on Data Streams

A rank or quantile query is to find a data entry with a given rank against a monotonic order specified on the data. Rank queries have several equivalent variations [107, 108, 109] and play important roles in many data stream applications.

It has been shown in [105] that an exact computation of rank queries requires memory size linear to the size of a data set by any one-scan technique, which may be impractical in on-line data stream computation where streams are massive in size and fast in arrival speed. Approximately computing rank queries over data streams has been investigated in the form of quantile computation.

A ϕ-quantile ($0 < \phi \leq 1$) of a collection of N data elements is the element with rank $\lceil \phi N \rceil$ against a monotonic order specified on the data set. The main paradigm is to continuously and efficiently maintain a small data structure (sketch/summary) in space over data elements for online queries. It has been shown in [110, 111, 112, 113] that a space-efficient ϕ-approximation quantile sketch can be maintained so that, for a quantile ϕ, it is always possible to find an element at rank r' with the uniform precision guarantee $\|r' - \lceil \phi N \rceil\| \leq \varepsilon N$. Due to the observation that many real data sets often exhibit skew towards heads (or tails depending on a given monotonic order), relative rank error (or biased) quantile computation techniques have been recently developed [114, 107, 109], which give better rank error guarantees towards heads.

Top-k queries have been extended to data streams. In [5], Mouratidis *el al.* study the problem of continuous monitoring top-k queries over sliding windows. Very recently, [115] improves the performance of the algorithms.

3.5.2.1 How Is Our Study Related?

All the existing studies on continuous ranking or quantile queries on data streams do not consider uncertain data. Those methods cannot be extended to probabilistic threshold top-k queries on uncertain data directly due to the complexity of possible worlds semantics. In this book, we investigate native methods for uncertain data streams.

3.5.3 Continuous Sensor Stream Monitoring

Sensor stream monitoring focuses on maintaining the answers to deterministic queries in sensor networks, while reducing the energy consumption as much as possible. Deshpande *et al.* [116] built a correlation-aware probabilistic model from the stored and current data in a sensor network, and use the probabilistic model to answer SQL queries. Only approximate answers with certain confidence bounds are provided, but the cost of data maintenance and query answering is significantly reduced. More specifically, Liu *et al.* [117] studied Min/Max query monitoring over distributed sensors. In their scenario, queries are submitted to a central server, and the major cost in query answering is the communication cost between the central server and distributed sensors. The authors model the reading of each sensor as a random variable, whose probability distribution can be obtained from historical data. Those distributions are used to estimate the answer to any Min/Max query. The server also contacts a small number of sensors for their exact readings, in order to satisfy the user specified error tolerance. [118] considers the applications where multiple sensors are deployed to monitor the same region. A sampling method is used to answer continuous probabilistic queries. The values of sensors that have little effect on the query results are sampled at a lower rate.

3.5.3.1 How Is Our Study Related?

There are three differences between our study and [116, 118, 117]. First, the uncertain data models adopted in the above work are different from the uncertain stream model discussed in this book, due to different application requirements. Second, the monitored queries are different: [116, 118, 117] deal with general SQL queries, Min/Max queries, and probabilistic queries, respectively, but our study focuses on top-k queries on uncertain streams specifically. Last, while the above work only provides approximate answers, our study can provide a spectrum of methods including an exact algorithm, a random method and their space efficient versions.

3.6 Probabilistic Linkage Queries

The problem of ranking queries on probabilistic linkages that will be discussed in Chapter 7 is mainly related to the existing work on record linkages and probabilistic graphical models.

3.6.1 Record Linkage

Computing record linkages has been studied extensively. Please refer to [37] as a nice tutorial. Generally, linkage methods can be partitioned into two categories. The deterministic record linkage methods [119] link two records if their values in certain matching attributes such as "name", "address" and "social insurance number" are exactly identical. The deterministic record linkage methods are not very effective in real-life applications due to data incompleteness and inconsistency.

Probabilistic record linkage methods [38] estimate the likelihood of two records being a match based on some similarity measures in the matching attributes. The similarity measures used in probabilistic record linkage methods fall into three classes [37].

- The first class is based on the Fellegi-Sunter theory [120]. The general idea is to model the values of the records in matching attributes as comparison vectors, and estimate the probability of two records being matched or unmatched given their comparison vectors [121, 122, 123].
- The second class is "edit based" measures such as the Levenshtein distance [124] and the edit distance [125].
- The third class is "term based" measures, where terms can be defined as words in matching attributes or Q-grams [126]. Such similarity measures include the fuzzy matching similarity [127] and the hybrids similarity measure developed in [128].

More recent work on record linkages in different scenarios includes [129, 130, 131, 132].

3.6.1.1 How Is Our Study Related?

In this book, we do not propose any record linkage methods. Instead, we focus on how to use probabilistic linkages produced by the existing probabilistic record linkage methods to answer aggregate queries in a meaningful and efficient way. As illustrated in Example 2.12, traditional post-processing methods that transform the probabilistic linkages into deterministic matches using thresholds may generate misleading results. Moreover, all existing record linkage methods only return linkage probabilities independently. There are no existing methods that output joint distributions. Therefore, deriving possible probabilities is far from trivial as will be shown in Chapter 7.

3.6.2 Probabilistic Graphical Models

Probabilistic graphical models refer to graphs describing dependencies among random variables. Generally, there are two types of probabilistic graphical models: directed graphical models [133] (also known as Bayesian networks or belief networks) and undirected graphical models [134] (also known as Markov networks or Markov random fields).

In directed graphical models, a vertex represents a random variable. A directed edge from vertex X_a to vertex X_b represents that the probability distribution of X_b is conditional on that of X_a.

In undirected graphical models, an edge between two random variables represents the dependency between the variables without particular directions. A random variable X_a is independent to all variables that are not adjacent to X_a conditional on all variables adjacent to X_a.

In an undirected graphical model, the joint probability distribution of the random variables can be factorized by the marginal distributions of the cliques in the graph, if the graph does not contain a loop of more than 3 vertices that is not contained in a clique [135].

3.6.2.1 How Is Our Study Related?

In this book, we develop PME-graphs as a specific type of undirected graphical models. We exploit the special properties of PME-graphs beyond the general undirected graphical models, and study the factorization of the joint probabilities in PME-graphs. Moreover, we develop efficient methods to evaluate aggregate queries on linkages using PME-graphs.

3.7 Probabilistic Path Queries

Last, we review the previous studies related to the probabilistic path queries that will be discussed in Chapter 8. The existing related work include path queries on probabilistic graphs and on traffic networks. Both problems have been studied extensively. However, there is no work on extending probabilistic graphs to traffic networks.

3.7.1 Path Queries on Probabilistic Graphs

Frank [136] studied the shortest path queries on probabilistic graphs, where the weight of each edge is a random variable following certain probability distribution. The probability distribution of the shortest path is defined. Moreover, a Monte Carlo simulation method is proposed to approximate the probability distribution of the shortest path. Loui [137] extended [136] by defining a utility function which specifies the preference among paths. Loui [137] also gave the computationally tractable formulations of the problem.

The shortest path problem on probabilistic graphs has been studied under different constraints. Hassin and Zemel [138] considered computing shortest paths when the edge weight distributions have a Taylor's series near zero. Wu *et al.* [139] studied the shortest path problem with edge weights uniformly and independently distributed in $[0,1]$. Moreover, Blei and Kaelbling [140] studied the problem of finding the shortest paths in stochastic graphs with partially unknown topologies. The problem is reduced to a Markov decision problem. Approximation algorithms are proposed. In [141], the problem of optimal paths in directed random networks is studied, where the cost of each arc is a real random variable following Gaussian distribution, and the optimal path is a path that maximizes the expected value of a utility function. In particular, linear, quadratic, and exponential utility functions are considered.

Stochastic shortest path search [142, 143, 144] is to find a path between two end nodes and maximize the probability that the path length does not exceed a given threshold. It is also referred to as the "stochastic on-time arrival problem (SOTA)". SOTA has the same semantics as the WT top-1 queries (a special case of WT top-*k* queries discussed in this book). However, Nikolova *et al.* [143] only consider some particular parametric weight distribution (such as the Normal distribution and the Poisson distribution) and transform the SOTA problem to a quasi-convex maximization problem. In addition, there have been other formulations of the optimal routing problem with probabilistic graphs. Ghosh *et al.* [145] developed an optimal routing algorithm that generates an optimal delivery subgraph so that the connectivity between two end nodes is maximized. Chang and Amir [146] computed the most reliable path between two end nodes when each edge has a failure probability.

Another related problem is traversing probabilistic graphs. Kahng [147] provided a nice overview and insights of this problem. Povan and Ball [148] showed that even approximating the probability that two vertices in a random graph are connected is NP-hard.

3.7.1.1 How Is Our Study Related?

Our work is different from the above studies in the following three aspects.

First, the probabilistic graph models are different. Many existing studies focus on simple probabilistic graphs, where probabilistic weights are independent from each other, such as [136, 144]. Moreover, some methods only work for certain probability distributions, such as uniform distribution [139] and the Normal distribution [143]. Last, some studies do consider correlations among edge weights. However, only certain types of correlations are considered, like the dependence with a global hidden variable [146]. Our model considers arbitrary weight distributions and correlations between the weights of adjacent edges. It is more capable and flexible for real road networks.

Second, the path queries are different. Most of the above studies focus on the optimal path query, where a utility function is adopted to evaluate the optimality of paths. However, using a single simple aggregate as the utility score may not capture traffic uncertainty sufficiently, since the probability of optimality is often very small. To tackle the problem, we propose probabilistic path queries and two top-k path queries.

Last, the query answering techniques are different. We propose a novel best-first search algorithm for probabilistic path queries. Moreover, we develop a hierarchical partition tree to index the road network structure and weight distribution information. Our query answering methods are efficient and scalable thanks to the two techniques.

3.7.2 Path Queries on Certain Traffic Networks

The shortest path queries on traffic networks have been studied extensively before. Please see [149] for a nice survey. The optimal algorithms often adopt dynamic programming and are not scalable for large scale traffic networks. As a result, heuristic algorithms that provide high quality approximate answers with short response time are developed.

The well known A* algorithm [150, 151] uses a heuristic evaluation function $f(x) = g(x) + h(x)$ to measure the optimality of the current explored path, where $g(x)$ is the cost from the start vertex to the current vertex, and $h(x)$ is the heuristic estimation of the distance to the goal. The paths with smaller $f(x)$ score are explored earlier.

Sanders and Schultes [152, 153] proposed a "highway hierarchy" for large scale traffic networks, which captures the key edges that are on the shortest paths between two far away vertices. Then, the shortest path search is restricted to those key edges.

Ertl [154] considered the geographical location of each edge in a traffic network and associated with each edge a radius, indicating how important the edge is in path search. An edge is only considered for a path if either the start vertex or the end vertex is inside the radius of the edge.

In [155], a hierarchical traffic network is proposed based on graph partitioning. When the shortest path search is far away from the start or end vertices, the algorithm only looks at the paths at higher levels of the hierarchical network.

In [156, 157], a concept of "transit" node is introduced to preprocess traffic networks. A transit node is on a set of non-local shortest paths. The distance from every vertex in the network to its closest transit node is computed to help the shortest path search.

In [158], important driving and speed patterns are mined from historical data, and are used to help to compute the fastest paths on traffic networks. A road hierarchy is built based on different classes of roads. Frequently traversed road segments are preferred in the path search.

In addition, Kurzhanski and Varaiya [159] considered a model that allows correlations between links for the reachability problem. More studies on the hierarchical approach for searching shortest path include [160, 161, 162, 163].

3.7.2.1 How Is Our Study Related?

The above studies tackle the path queries in large scale certain traffic networks. Therefore, both the graph models and the query types are different from our work. Thus, those techniques cannot be extended to probabilistic path queries on uncertain road networks.

Moreover, although hierarchical indices have been extensively used in path queries on certain traffic networks, the existing index techniques only work for certain path queries. Thus, we develop a hierarchical partition tree to index the weight probability distributions on graphs.

Chapter 4
Top-k Typicality Queries on Uncertain Data

An uncertain object O can be modeled as a set of instances generated by an underlying random variable $\mathscr{X}$. *If there are a large number of instances in an uncertain object, how can we understand and analyze this object?* An effective way is to find the most typical instances among all instances of the uncertain object. In Section 2.2.1, we applied the idea of typicality analysis from psychology and cognitive science to ranking uncertain data, and modeled typicality for instances in uncertain objects systematically. Three types of top-k typicality queries are formulated.

- To answer questions such as "*Who are the top-k most typical NBA players?*", the measure of *simple typicality* is developed.
- To answer questions like "*Who are the top-k most typical guards distinguishing guards from other players?*", the notion of *discriminative typicality* is proposed.
- To answer questions like "*Who are the best k typical guards in whole representing different types of guards?*", the notion of *representative typicality* is used.

Computing the exact answer to a top-k typicality query requires quadratic time which is often too costly for online query answering on uncertain objects with large number of instances. In this chapter, we develop a series of approximation methods for various situations.

- The randomized tournament algorithm has linear complexity though it does not provide a theoretical guarantee on the quality of the answers.
- The direct local typicality approximation using VP-trees provides an approximation quality guarantee.
- A Local Typicality Tree data structure can be exploited to index a large set of instances. Then, typicality queries can be answered efficiently with quality guarantees by a tournament method based on a Local Typicality Tree.

An extensive performance study using two real data sets and a series of synthetic data sets clearly shows that top-k typicality queries are meaningful and our methods are practical.

4.1 Answering Simple Typicality Queries

Consider an uncertain object O, the simple typicality value of an instance $o \in O$ is the likelihood of o given that o is a sample of $\mathscr{X}$, the underlying random variable generating the samples in O (Definition 2.5).

In this section, we first discuss how to compute likelihood values, then, we show that the complexity of answering top-k typicality queries is quadratic. Last, we present a randomized tournament approximation algorithm (RT). The approximation algorithm developed for simple typicality computation in this section can be extended to answer top-k discriminative typicality queries and top-k representative typicality queries, as will be discussed later in Sections 4.3 and 4.4, respectively.

4.1.1 Likelihood Computation

For an instance o in an uncertain object O, likelihood $L(o|O)$ is the posterior probability of o given instances in O, which can be computed using probability density estimation methods. There are several model estimation techniques in the literature [164], including parametric and non-parametric density estimation. Parametric density estimation requires a certain distribution assumption, while non-parametric estimation does not. Among the various techniques proposed for non-parametric density estimation [165], histogram estimation [166], kernel estimation [167, 168] and nearest neighbor estimation [169] are the most popular. In this work, we use kernel estimation, because it can estimate unknown data distributions effectively [170].

Kernel estimation is a generalization of sampling. In random sampling, each sample point carries a unit weight. However, an observation of the sample point increases the chance of observing other points nearby. Therefore, kernel estimator distributes the weight of each point in the nearby space around according to a *kernel function K*. The commonly used kernel functions are listed in Table 4.1, where $I(|u| \leq 1)$ in the kernel functions denotes the value 1 when $|u| \leq 1$ holds, and 0 when $|u| \leq 1$ does not hold.

A bandwidth parameter (also known as the smoothing parameter) h is used to control the distribution among the neighborhood of the sample. As shown in [171], the quality of the kernel estimation depends mostly on the bandwidth h and lightly on the choice of the kernel K. Too small bandwidth values cause very spiky curves, and too large bandwidth values smooth out details. A class of effective methods are data-driven least-squares cross-validation algorithms [172, 173, 174, 175], which select the bandwidth value that minimizes integrated square error.

In this work, we choose the commonly used Gaussian kernels. Our approach can also be adapted to using other kernel functions. We set the bandwidth of the Gaussian kernel estimator $h = \frac{1.06s}{\sqrt[5]{n}}$ as suggested in [175], where n is the cardinality of the uncertain object O and s is the standard deviation of the instances in O which can be estimated by sampling. In Section 4.5.3, we evaluate the sensitivity of the answers to top-k typicality queries with respect to the choice of kernel functions and bandwidth values. The results show that the answers computed using different kernel functions listed in Table 4.1 are mostly consistent. Moreover, using different bandwidth values around $h = \frac{1.06s}{\sqrt[5]{n}}$ also provide consistent answers.

Outliers in instances may increase the standard deviation of the instances in O, and thus lead to larger bandwidth values, which may impair the quality of the answers to typicality queries. Therefore, for better performance, we can remove outliers in the set of instances as preprocessing. There are extensive studies on effective and efficient outlier detection [176, 177, 178], which can be used as a screening step in our methods. Moreover, it is shown from the experimental results in Section 4.5.3 that, even on an uncertain object containing a non-trivial amount of noise, the results returned by top-k typicality queries are often consistent with the results found when outliers are removed.

Name	Kernel function
Uniform	$K(u) = \frac{1}{2}I(\lvert u\rvert \leq 1)$
Triangle	$K(u) = (1-\lvert u\rvert)I(\lvert u\rvert \leq 1)$
Epanechnikov	$K(u) = \frac{3}{4}(1-u^2)I(\lvert u\rvert \leq 1)$
Quartic	$K(u) = \frac{15}{16}(1-u^2)^2 I(\lvert u\rvert \leq 1)$
Triweight	$K(u) = \frac{35}{32}(1-u^2)^3 I(\lvert u\rvert \leq 1)$
Gaussian	$K(u) = \frac{1}{\sqrt{2\pi}}e^{-\frac{1}{2}u^2}$
Cosine	$K(u) = \frac{\pi}{4}\cos(\frac{\pi}{2}u)I(\lvert u\rvert \leq 1)$

Table 4.1 The commonly used kernel functions.

Since we address the top-k typicality problem in a generic metric space, the only parameter we use in density estimation is the distance (or similarity) between two instances. Formally, given an uncertain object $O = (o_1, o_2, \ldots, o_n)$ in a generic metric space, the underlying likelihood function is approximated as

$$L(x|O) = \frac{1}{nh}\sum_{i=1}^{n} G_h(x, o_i) = \frac{1}{nh\sqrt{2\pi}}\sum_{i=1}^{n} e^{-\frac{d(x,o_i)^2}{2h^2}} \tag{4.1}$$

where $d(x, o_i)$ is the distance between x and o_i in the metric space, and $G_h(x, o_i) = \frac{1}{\sqrt{2\pi}}e^{-\frac{d(x,o_i)^2}{2h^2}}$ is a *Gaussian kernel*.

Hereafter, by default, we assume that outliers are removed using the techniques discussed in Section 4.1.1.

4.1.2 An Exact Algorithm and Complexity

Theoretically, given an uncertain object O, if the likelihood of an instance $o \in O$ satisfies $L(o|O) \propto \frac{1}{\sum_{o' \in O} d(o, o')}$, then the discrete 1-median problem can be reduced

Algorithm 4.1 ExactTyp(OS,k)

Input: an uncertain object $O = \{o_1, \dots, o_n\}$ and positive integer k
Output: the k instances with the highest simple typicality values
Method:
1: **for all** instance $o \in O$ **do**
2: set $T(o,O) = 0$
3: **end for**
4: **for** $i = 1$ to n **do**
5: **for** $j = i+1$ to n **do**
6: $w = \frac{1}{nh\sqrt{2\pi}} e^{-\frac{d(o_i,o_j)^2}{2h^2}}$
7: $T(o_i,O) = T(o_i,O) + w$
8: $T(o_j,O) = T(o_j,O) + w$
9: **end for**
10: **end for**
11: return the top-k instances according to $T(o,O)$

to a special case of the top-1 simple typicality query problem. As so far no better than quadratic algorithm has been found for exact solutions to the general discrete 1-median problem (except in L_1 metric space), it is challenging to find a better than quadratic algorithm for computing exact answers to general top-k typicality queries.

We now present Algorithm 4.1, a straightforward method that computes the exact answer to a top-k simple typicality query. It computes the exact simple typicality for each instance using two nested loops, and then selects the top-k instance. The complexity of Algorithm 4.1 is $O(|O|^2)$, where $|O|$ is the number of instances in O. Quadratic algorithms are often too costly for online queries on large databases, while good approximations of exact answers are good enough for typicality analysis. This motivates our development of approximation algorithms.

4.1.3 A Randomized Tournament Algorithm

Inspired by the randomized tournament method [78] for the discrete 1-median problem, we propose a randomized tournament algorithm for answering top-k simple typicality queries as follows.

Let t be a small integer, called the *tournament group size*. To find the most typical instances in object O of n instances, we partition the instances into $\lceil \frac{n}{t} \rceil$ groups randomly such that each group has t instances. For each group, we use the exact algorithm to find the instance that has the largest simple typicality value in the group. Only the winner instances in the groups are sent to the next round.

The winners of the previous round are again partitioned randomly into groups such that each group contains t instances. The most typical instance in each group is selected and sent to the next round. The tournament continues until only one instance is selected as the winner. The final winner is an approximation of the most typical instance.

Algorithm 4.2 RandomTyp(O,k,t,v)

Input: an uncertain object O, positive integer k, tournament size t and number of validations v
Output: approximation to the answer to a top-k simple typicality query $\widetilde{A}$
Method:

```
1: Ã = ∅
2: for i = 1 to k do
3:    O' = O − Ã
4:    candidate = ∅
5:    for j = 1 to v do
6:       repeat
7:          G = {g_i} (1 ≤ i ≤ ⌈n/t⌉, |g_i| = t, ∪_{g_i∈G} g_i = S')
8:          for all group g ∈ G do
9:             winner_g = ExactTyp(g, 1)
10:            O' = O' − g ∪ winner_g
11:         end for
12:      until |O'| = 1
13:      candidate = candidate ∪ O'
14:   end for
15:   Ã = Ã ∪ {argmax_{o∈candidate}{T(o, O)}}
16: end for
```

To approximate the second most typical instance, we run the randomized tournament again with the following constraint: the most typical instance already chosen in the previous tournament cannot be selected as the winner in this tournament. The final winner in the second tournament is the approximation of the second most typical instance. Continuing in this manner, we can find an approximation to the set of top-k typical instances by running a total of k tournaments.

In order to achieve a higher accuracy, we can run this randomized tournament several times for selecting the approximation of the i-th most typical instance ($1 \leq i \leq k$), and pick the instance with the largest simple typicality among all the final winners. The procedure is given in Algorithm 4.2.

The typicality computation within one group has the time complexity of $O(t^2)$. There are $\lceil \log_t n \rceil$ tournament rounds in total. Without loss of generality, let us assume $n = t^m$. Then, the first round has $\frac{n}{t}$ groups, the second round has $\frac{\frac{n}{t}}{t} = \frac{n}{t^2}$ groups, and so forth. The total number of groups is $\sum_{1 \leq i \leq \log_t n} \frac{n}{t^i} = \frac{n}{t-1}(1 - \frac{1}{t^m}) = O(\frac{n}{t})$. The complexity of selecting the final winner is $O(t^2 \cdot \frac{n}{t}) = O(tn)$. If we run each tournament v times for better accuracy, and run tournaments to choose top-k typical instances, the overall complexity is $O(kvtn)$.

The randomized algorithm runs in linear time with respect to the number of instances. However, the accuracy of the approximation to the answer is not guaranteed in theory, though in practice it often has reasonable performance.

4.2 Local Typicality Approximation

While the randomized tournament method is efficient, it does not have formally provable accuracy. Can we provide some quality guarantee and at the same time largely retain the efficiency? In this section, we develop several heuristic local typicality approximation methods. Our discussion in this section is for simple typicality. The methods will be extended to other typicality measures later in this chapter.

4.2.1 Locality of Typicality Approximation

In Gaussian kernel estimation, given two instances a and p in an uncertain object O, the contribution from p to $T(a,S)$, the simple typicality score of a, is $\frac{1}{nh\sqrt{2\pi}}e^{-\frac{d(a,p)^2}{2h^2}}$, where n is the cardinality of the uncertain object O. The contribution of p decays exponentially as the distance between a and p increases. Therefore, if p is remote from a, p contributes very little to the simple typicality score of a.

Moreover, in a metric space, given three instances a, b and p, the triangle inequality $|d(a,p)-d(b,p)| < d(a,b)$ holds. If $d(a,p) \gg d(a,b)$, then $d(a,p) \approx d(b,p)$. Therefore, the instances far away from a and b will have similar contributions to the probability density values $T(a,O)$ and $T(b,O)$.

Based on the above observations, given an uncertain object O and a subset $C \subseteq O$, can we use the locality to approximate the instance having the largest simple typicality value in C?

Definition 4.1 (Neighborhood region). Given an uncertain object O, a neighborhood threshold σ, and a subset $C \subseteq O$, let $D = D_{A_1} \times \cdots \times D_{A_n}$ where D_{A_i} is the domain of attribute A_i ($1 \leq i \leq n$), the **σ-neighborhood region** of C is defined as $D(C,\sigma) = \{o | o \in D, \min_{o' \in C}\{d(o,o')\} \leq \sigma\}$. ■

Definition 4.2 (Local simple typicality). Given an uncertain object O, a neighborhood threshold σ, and a subset $C \subseteq O$, let $\mathscr{X}$ be the random vector that generates samples O, the **local simple typicality** of an object $o \in C$ is defined as $LT(o,C,\mathscr{X},\sigma) = L(o|\mathscr{X}_{D(C,\sigma)})$ where $L(o|\mathscr{X}_{D(C,\sigma)})$ is the likelihood of o given that it is a sample of $\mathscr{X}$ in region $D(C,\sigma)$. ■

In practice, for each instance $o \in O$, we use the set of instances in O that lie in o's σ-neighborhood region to estimate the simple typicality of o.

Definition 4.3 (Local neighborhood). Given an uncertain object O, a neighborhood threshold σ, and a subset $C \subseteq O$, The **σ-neighborhood** of C is defined as $LN(C,O,\sigma) = \{o | o \in O \cap D(C,\sigma)\}$, where $D(C,\sigma)$ is the σ-neighborhood region of C. ■

$LN(C,O,\sigma)$ is the set of instances in O whose distance to at least one instance in C is at most σ. Then, $LT(o,C,\mathscr{X},\sigma)$ can be estimated using $LT(o,C,O,\sigma) =$

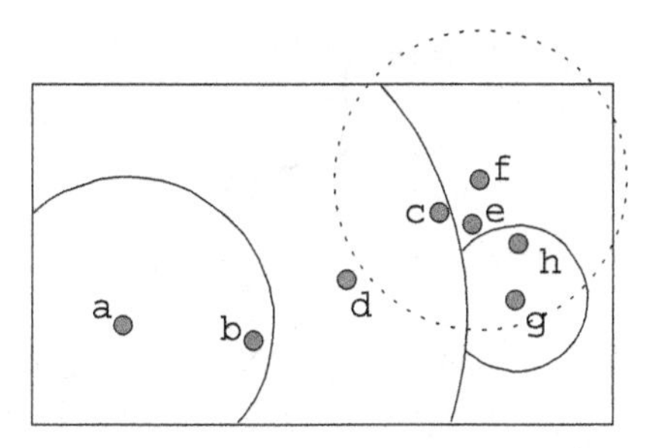

Fig. 4.1 Decomposing a set of instances in a VP-tree.

$L(o|LN(o,C,\sigma))$, where $L(o|LN(o,C,\sigma)$ is the likelihood of o given objects $LN(o,C,\sigma)$.

The following result uses local simple typicality to approximate the simple typicality with a quality guarantee.

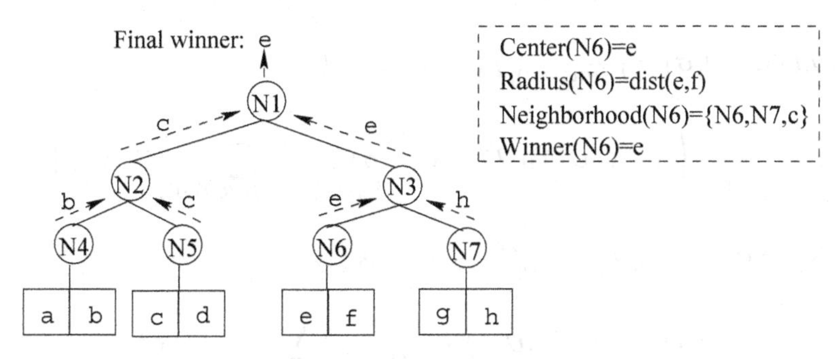

Fig. 4.2 Computing the approximate most typical instance.

Theorem 4.1 (Local typicality approximation). *Given an uncertain object O, neighborhood threshold σ, and a subset $C \subseteq O$, let $\widetilde{o} = \arg\max_{o_1 \in C}\{LT(o_1, C, O, \sigma)\}$ be the instance in C having the largest local simple typicality value, and $o = \arg\max_{o_2 \in C}\{T(o_2, O)\}$ be the instance in C having the largest simple typicality value. Then,*

$$T(o,O) - T(\widetilde{o},O) \leq \frac{1}{h\sqrt{2\pi}} e^{-\frac{\sigma^2}{2h^2}} \tag{4.2}$$

Moreover, for any object $x \in C$,

$$T(x,O) - LT(x,C,O,\sigma) < \frac{1}{h\sqrt{2\pi}} e^{-\frac{\sigma^2}{2h^2}} \leq \frac{1}{h\sqrt{2\pi}} \tag{4.3}$$

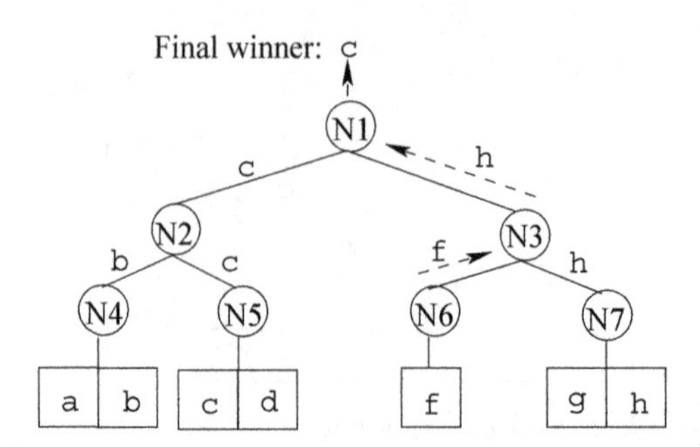

Fig. 4.3 Computing the approximate second most typical instance.

Proof. For any instance $x \in C$,

$$T(x,O) = \frac{1}{|S|h}\left(\sum_{y \in LN(C,S,\sigma)} G_h(x,y) + \sum_{z \in (S-LN(C,O,\sigma))} G_h(x,z)\right)$$

Since $LT(x,C,O,\sigma) = \frac{1}{h|LN(C,O,\sigma)|}\sum_{y \in LN(C,O,\sigma)} G_h(x,y)$,

$$T(x,O) = \frac{1}{h|O|}\left(|LN(C,O,\sigma)| \cdot LT(x,C,O,\sigma) + \sum_{z \in (O-LN(C,O,\sigma))} G_h(x,z)\right) \quad (4.4)$$

Because $LN(C,O,\sigma) \subseteq O$, $\frac{|LN(C,O,\sigma)|}{|O|} \leq 1$. Thus,

$$T(x,O) \leq LT(x,C,O,\sigma) + \frac{1}{h|O|}\sum_{z \in (O-LN(C,O,\sigma))} G_h(x,z) \quad (4.5)$$

According to the definition of local neighborhood, $d(x,z) > \sigma$ for any $z \in (O - LN(C,O,\sigma))$. Thus,

$$\frac{1}{|O|}\sum_{y \in (O-LN(C,O,\sigma))} G_h(x,y) < \frac{1}{\sqrt{2\pi}} e^{-\frac{\sigma^2}{2h^2}} \quad (4.6)$$

Inequality 4.3 follows from Inequalities 4.5 and 4.6 immediately. Applying Equation 4.4 to o and $\widetilde{o}$, respectively, we have

$$\begin{aligned} T(o,O) - T(\widetilde{o},O) = & \frac{|LN(C,O,\sigma)|}{|O|}(LT(o,C,O,\sigma) - LT(\widetilde{o},C,O,\sigma)) \\ & + \frac{1}{h|O|}\sum_{z \in (O-LN(C,O,\sigma))}(G_h(o,z) - G_h(\widetilde{o},z)) \end{aligned}$$

Using Inequality 4.6, we have

$$\frac{1}{|O|}\sum_{z\in(O-LN(C,O,\sigma))}(G_h(o,z)-G_h(\widetilde{o},z))\leq\frac{1}{\sqrt{2\pi}}e^{-\frac{\sigma^2}{2h^2}}$$

Since $LT(\widetilde{o},C,O,\sigma)\geq LT(o,C,O,\sigma)$, $LT(o,C,O,\sigma)-LT(\widetilde{o},C,O,\sigma)\leq 0$. Thus,

$$T(o,O)-T(\widetilde{o},O)\leq\frac{1}{h\sqrt{2\pi}}e^{-\frac{\sigma^2}{2h^2}}$$

Inequality 4.2 is shown. ■

From Theorem 4.1, we can also derive the minimum neighborhood threshold value σ to satisfy certain approximation quality.

Corollary 4.1 (Choosing neighborhood threshold). *Given an uncertain object O, an instance $x\in S$, and a quality requirement θ, if $\sigma\geq\sqrt{-2\ln\sqrt{2\pi}\theta h}\cdot h$, then $T(x,O)-LT(x,C,O,\sigma)<\theta$ for any subset C ($x\in C$).*

Proof. From Theorem 4.1, for any instance $o\in C$, subset $C\subseteq O$ and neighborhood threshold σ, we have

$$T(x,O)-LT(x,C,O,\sigma)<\frac{1}{h\sqrt{2\pi}}e^{-\frac{\sigma^2}{2h^2}}$$

In order to meet the quality requirement θ, it should hold that $\frac{1}{h\sqrt{2\pi}}e^{-\frac{\sigma^2}{2h^2}}\leq\theta$. Therefore, $\sigma\geq\sqrt{-2\ln\sqrt{2\pi}\theta h}\cdot h$. ■

4.2.2 DLTA: Direct Local Typicality Approximation Using VP-trees

Inequality 4.3 in Theorem 4.1 can be used immediately to approximate simple typicality computation with quality guarantee. Given a neighborhood threshold σ, for each instance $x\in S$, the direct local typicality approximation (DLTA) algorithm computes the σ-neighborhood of $\{x\}$, i.e., $LN(\{x\},O,\sigma)$ and the local simple typicality $LT(x,\{x\},O,\sigma)$. Then, it returns the k instances with the highest local simple typicality values as the approximation to the answer of the top-k simple typicality query.

The quality of the approximation answer is guaranteed by the following theorem.

Theorem 4.2 (Approximation quality). *Given an uncertain object O, neighborhood threshold σ and integer k, let A be the k instances with the highest simple typicality values, and $\widetilde{A}$ be the k instances with the highest local simple typicality values. Then,*

$$\frac{\sum_{o\in A}T(o,O)-\sum_{\widetilde{o}\in\widetilde{A}}T(\widetilde{o},O)}{k}<\frac{1}{h\sqrt{2\pi}}e^{-\frac{\sigma^2}{2h^2}} \tag{4.7}$$

Proof. If $A \cap \widetilde{A} \neq \emptyset$, then let $A = A - A \cap \widetilde{A}$ and $\widetilde{A} = \widetilde{A} - A \cap \widetilde{A}$. So in the rest of the proof, we assume $A \cap \widetilde{A} = \emptyset$.

We sort the instances in A in the descending order of their typicality values, and sort the instances in $\widetilde{A}$ in the descending order of their local typicality values. Let o be the i-th instance in A, and $\widetilde{o}$ be the i-th instance in $\widetilde{A}$ ($1 \leq i \leq k$). From Inequality 4.3, we have

$$0 \leq T(o,O) - LT(o,\{o\},O,\sigma) < \frac{1}{h\sqrt{2\pi}} e^{-\frac{\sigma^2}{2h^2}}$$

and

$$0 \leq T(\widetilde{o},O) - LT(\widetilde{o},\{\widetilde{o}\},O,\sigma) < \frac{1}{h\sqrt{2\pi}} e^{-\frac{\sigma^2}{2h^2}}$$

Moreover, since $o \notin \widetilde{A}$, it holds that $LT(o,\{o\},O,\sigma) < LT(\widetilde{o},\{\widetilde{o}\},O,\sigma)$. Thus,

$$\begin{array}{l} T(o,O) - T(\widetilde{o},S) = (T(o,O) - LT(o,\{o\},O,\sigma)) \\ -(T(\widetilde{o},O) - LT(\widetilde{o},\{\widetilde{o}\},O,\sigma)) + (LT(o,\{o\},O,\sigma) \\ -LT(\widetilde{o},\{\widetilde{o}\},O,\sigma)) < \frac{1}{h\sqrt{2\pi}} e^{-\frac{\sigma^2}{2h^2}} \end{array}$$

Inequality 4.7 follows by summing up the above difference at each rank i ($1 \leq i \leq k$). ■

Searching the σ-neighborhood for each instance can be very costly. To implement the direct local typicality approximation efficiently, we can use the VP-tree index [179] to support σ-neighborhood searches effectively.

A VP-tree [179] is a binary space partitioning (BSP) tree. Given a set of instances O, a VP-tree T indexes the instances in O. Each node in a VP-tree represents a subset of O. Roughly speaking, for each non-leaf node N and the set of nodes O_N at N, a *vantage point* is selected to divide the set O_N into two exclusive subsets O_{N_1} and O_{N_2} ($O_N = O_{N_1} \cup O_{N_2}$) such that, to search the instances within distance σ to an instance $p \in N$, likely we only need to search either O_{N_1} or O_{N_2} but not both. O_{N_1} and O_{N_2} are used to construct the two children of N. For example, the VP-tree in Figure 4.2 indexes the instances in Figure 4.1.

A VP-tree can be constructed top-down starting from the root which represents the whole set of instances. A sampling method is given in [179] to select vantage points for internal nodes. Then, the first half subset of instances that are close to the vantage point form the left child of the root, and the other instances form the right child. The left and the right children are further divided recursively until a node contains only one instance (a leaf node). A VP-tree can be constructed in cost $O(|O| \log |O|)$.

Searching a VP-tree for the σ-neighborhood of a query point is straightforward using the recursive tree search. Once an internal node in the tree can be determined in the σ-neighborhood of the query point, all descendant instances of the internal node are in the neighborhood and no subtrees need to be searched. For example, the σ-neighborhood of node $N_6 = \{e, f\}$ in Figure 4.1 is represented by the dashed circle. To find all points in the σ-neighborhood of N_6, we search the VP-tree in Figure 4.2 from the root node N_1, and recursively examine each internal node. During the search, node N_4 can be pruned since all points in N_4 lie out of the σ-neighborhood of N_6.

The cost of computing the local simple typicality of an instance x is $O(|LN(\{x\},O,\sigma)|)$. Then, computing the local simple typicality of all instances in O takes $O(\sum_{x\in O}|LN(\{x\},O,\sigma)|)$ time. Although the search can be sped up using a VP-tree, the complexity of the DLTA algorithm is still $O(|O|^2)$. The reason is, in the worst case where σ is larger than the diameter (i.e., the largest pairwise distance) of the data set, the σ-neighborhood of each instance contains all other instances in O.

4.2.3 LT3: Local Typicality Approximation Using Tournaments

To reduce the cost of computing local simple typicality further, we incorporate the tournament mechanism, and propose a local typicality approximation algorithm using tournaments (the LT3 algorithm). The basic idea is to group the instances locally, and conduct a tournament in each local group of instances. The instances with the largest local simple typicality is selected as the winner. The winners are sent to the next round of tournament. The tournaments terminate when only one instance is left as the winner. A sampling method is employed to reduce the computational cost.

4.2.3.1 Local Typicality Trees (LT-trees)

A local typicality tree (LT-tree) is an MVP-tree [180] with auxiliary information that supports local typicality calculation and tournaments. Given an uncertain object O, an LT-tree can be constructed as follows.

First, we construct an MVP-tree [180] on O, which is a t-nary VP-tree that uses more than one vantage point to partition the space. Without loss of generality, let us assume $t = 2^l$ and the data set contains t^m instances. We assign a layer number to each node in the MVP-tree. The root node has layer number 0, and a node is assigned layer number $(i+1)$ if its parent has layer number i. We remove all those nodes in the MVP-tree whose layer number is not a multiple of t. For a node N of layer number jt $(j \geq 1)$, we connect N to its ancestor in the MVP-tree of layer $(j-1)t$.

Second, we compute three pieces of information, the approximate center, the radius, and the σ-neighborhood, for each node in the LT-tree.

Approximate Center

For a node N in the LT-tree, let O_N be the set of instances at N. To compute the approximate center at a node N in the LT-tree, we draw a sample R of $\sqrt{|O_N|}$ instances from O_N, and compute the pairwise distance between every two instances in

R. Then, for each instance $x \in R$, the center-score of x is the maximum distance from x to another point in R. The instance in R of the minimum center-score is chosen as the center. This center approximation procedure is popularly used in computational geometry. It takes $O(|O_N|)$ time for each node N, and $O(|O|\log_t |O|)$ time for all nodes in the LT-tree.

Radius

Once the center c of a node N is chosen, the radius is given by the maximum distance between c and the other instances at N. This can be computed in time $O(|O_N|)$ for each node N, and $O(|O|\log_t |O|)$ for all nodes in the LT-tree.

σ-neighborhood

We use a range query in the LT-tree to compute a superset of the σ-neighborhood of O_N for every node N in the LT-tree, which always achieves a typicality approximation no worse than using the σ-neighborhood. To compute the superset, we start from the root and iteratively search for the nodes that completely lie in the σ-neighborhood of N, using the approximate center and radius of N. Once all objects at a node N' are in the σ-neighborhood of N, we use N' to represent them and do not search any subtrees of N'.

4.2.3.2 Query Answering

To answer a top-k simple typicality query, we run tournaments on the LT-tree in a bottom-up fashion. First, a tournament is run for each leaf node in the LT-tree. The winner enters the tournament at the parent node of the leaf node. The winner o_1 at the root node is the approximation of the most typical instance. Figure 4.2 illustrates the procedure of computing the approximate most typical point in the set of points in Figure 4.1 using an LT-tree. During the tournaments in the leaf nodes, $\{b, c, e, h\}$ are selected as local winners and sent to the parent nodes. Then, $\{c, e\}$ are selected in the tournaments in nodes N_2 and N_3. Finally, e is selected as the winner, which approximates the most typical point in the data set.

To find the approximation of the second most typical instance, we do not need to completely run the tournaments again. Instead, we can reuse most of the results in the tournaments of finding the most typical instance w_1. The only tournaments we need to rerun are on the nodes containing w_1. First, we run a new tournament at the leaf node N containing w_1, but do not include w_1 in the new tournament. Then, the winner w_1' is sent to N_p, the parent of N, and a new tournament is run there by replacing w_1 by w_1'. A series of m tournaments are needed to find a new winner w_2 in the root node, which is the approximation of the second most typical instance. At each level of the LT-tree, only one node needs to run a tournament. For example, in

Algorithm 4.3 LT3Typ(O,k,T)

Input: an uncertain object $O = \{o_1, \ldots, o_n\}$ and positive integer k and an LT-tree T (with m levels) built on O whose root node is N_R
Output: approximation to the answer to a top-k simple typicality query $\widetilde{A}$
Method:

1: $\widetilde{A} = \emptyset$
2: **for** $j = m$ to 0 **do**
3: **for all** node N at level L_j **do**
4: $winner_N = \arg\max_{o \in N}\{LT(o,N,O,\sigma)\}$
5: $N_p = N_p \cup \{winner_N\}$
$\{*N_p$ is the parent node of $N\}$
6: **end for**
7: **end for**
8: $\widetilde{A} = \widetilde{A} \cup \{winner_{N_R}\}$
9: $w_1 = winner_{N_R}$
10: **for** $i = 2$ to k **do**
11: find N such that $w_{i-1} \in N$
$\{*w_{i-1}$ is the last output winner$\}$
12: **while** $N \neq N_R$ **do**
13: $winner_N = \arg\max_{o \in N, o \neq w_{i-1}}\{LT(o,N,O,\sigma)\}$
14: $N_p = N_p \cup \{winner_N\}$
15: $N \leftarrow N_p$
16: **end while**
17: $\widetilde{A} = \widetilde{A} \cup \{winner_{N_R}\}$
18: $w_i = winner_{N_R}$
19: **end for**

the LT-tree in Figure 4.3, after e is selected as the first final winner, it is removed from node N_6. Then, only the tournaments in nodes N_6, N_3 and N_1 need to be reconducted. Finally, c is selected as the final winner to approximate the second most typical point. The complete procedure is shown in Algorithm 4.3.

The quality of the answers returned by the LT3 algorithm can be guaranteed by the following theorem.

Theorem 4.3. *In an uncertain object O, let o be the instance with the largest simple typicality and $\widetilde{o}$ be an instance computed by the LT3 method using local typicality approximation and the tournament group size t. Then,*

$$T(o,O) - T(\widetilde{o},O) < \left(\frac{1}{h\sqrt{2\pi}} e^{-\frac{\sigma^2}{2h^2}}\right) \cdot \lceil \log_t |O| \rceil$$

Proof. In the worst case, instance o is not selected as the winner in the first level of tournaments.

Let o_1 be the winner of the group containing o in the first level of tournaments, then we have

$$T(o,O) - T(o_1,O) < \frac{1}{h\sqrt{2\pi}} e^{-\frac{\sigma^2}{2h^2}}$$

as indicated by Inequality 4.2 in Theorem 4.1.

If o_i fails the $(i+1)$-th level of tournaments, let o_{i+1} be the winner of the group containing o_i in this tournament, then again we have

$$T(o_i, O) - T(o_{i+1}, O) < \frac{1}{h\sqrt{2\pi}} e^{-\frac{\sigma^2}{2h^2}}$$

For a set of instances in O, there are $\lceil \log_t |O| \rceil$ levels of tournaments. The final winner $\widetilde{o}$ is the winner in the $\lceil \log_t |O| \rceil$-th level of tournaments. That is, $\widetilde{o} = o_{\lceil \log_t |O| \rceil}$. Then,

$$\begin{aligned} T(o,O) - T(\widetilde{o},O) &= T(o,O) - T(o_1,O) + T(o_1,O) - T(o_2,O) + \dots \\ &+ T(o_{\lceil \log_t |O| \rceil - 1}, O) - T(o_{\lceil \log_t |O| \rceil}, O) < \frac{1}{h\sqrt{2\pi}} e^{-\frac{\sigma^2}{2h^2}} \cdot \lceil \log_t |O| \rceil \end{aligned}$$

The inequality in the theorem holds. ■

The LT3 algorithm combines the merits of both local typicality approximation and the tournament mechanism. It achieves better accuracy than the randomized tournament algorithm, thanks to the local grouping. It is more efficient than the DLTA algorithm because of the tournament mechanism. As shown in our experiments, the approximations of the most typical instances computed by the LT3 algorithm are very close to the exact ones. LT3 is very efficient and scalable.

4.2.3.3 A Sampling Method for Bounding Runtime

To make the analysis complete, here we provide a sampling method which provides an upper bound on the cost of local typicality computation with quality guarantee.

Suppose we want to compute the local simple typicality $LT(p, C, O, \sigma)$. We consider the contribution of an object $o \in LN(C, O, \sigma)$ to $LT(p, C, O, \sigma)$, denoted by

$$\eta(o) = \frac{1}{h|LN(C,O,\sigma)|} G_h(p,o) = \frac{e^{-\frac{d(p,o)^2}{2h^2}}}{h|LN(C,O,\sigma)|\sqrt{2\pi}}$$

We can draw a sample of $LN(C, O, \sigma)$ to estimate the expectation of $\eta(o)$. Please note that

$$LT(p,C,O,\sigma) = |LN(C,O,\sigma)| \cdot E[\eta(o)]$$

where $E[\eta(o)]$ is the expectation of $\eta(o)$ for $o \in LN(C, O, \sigma)$.

Using the Chernoff-Hoeffding bound [181], we derive the minimum sample size to achieve the required accuracy.

Theorem 4.4. *For any δ $(0 < \delta < 1)$ and ε $(\varepsilon > 0)$ and a sample R of $LN(C, O, \sigma)$, if $|R| > \frac{3h\sqrt{2\pi} \cdot e^{\frac{\sigma^2}{2h^2}} \cdot \ln \frac{2}{\delta}}{\varepsilon^2}$, then*

$$Pr\{|\frac{|LN(C,O,\sigma)|}{|R|} \sum_{o \in R} \eta(o) - LT(p,C,O,\sigma)| > \varepsilon \cdot LT(p,C,O,\sigma)\} < \delta$$

Proof. The theorem can be proved directly using a special form of the Chernoff-Hoeffding bound due to Angluin and Valiant [181]. ∎

Theorem 4.4 provides an upper bound of the sample size, which is independent of the size of data sets. The larger ε and δ, the smaller the sample size. The larger σ, the larger the sample size.

Using the sampling method suggested by Theorem 4.4, we can have a tournament algorithm using an LT-tree of cost $O(|O|\log|O|)$. The algorithm provides a theoretical bound on the runtime.

However, the sampling method cannot be practically gainful unless on extremely large data sets. The LT-tree already exploits the locality of instances nicely. When the data set is not extremely large, the number of instances in the σ-neighborhood of a node is usually (substantially) smaller than the number of samples required for high approximation quality. In our experiments, the above case is always true. Thus, we do not include the experimental results on this sampling method.

4.3 Answering Discriminative Typicality Queries

According to Definition 2.6, discriminative typicality can be calculated as follows. Given two uncertain objects O and S, where O is the target uncertain object, for each instance $o \in O$, Algorithm 4.1 can be used to compute the simple typicality scores of o in O and S, respectively. The difference between the two is the discriminative typicality of o.

Suppose there are m instances in the target object O and n instances in S. To compute the discriminative typicality score of an instance $o \in O$, we have to compute the simple typicality scores of o in both O and S, which takes $O(n+m)$ time. Therefore, answering top-k discriminative typicality queries using the exact algorithm takes time $O(m(m+n))$.

The approximation methods developed for top-k simple typicality queries can also be adopted to answer top-k discriminative typicality queries.

4.3.1 A Randomized Tournament Algorithm

Generally, the randomized tournament algorithm can be used to answer top-k discriminative typicality queries if the discriminative typicality measure is applied. In contrast to the randomized tournament algorithm for top-k simple typicality queries, only the instances in the target object O are involved in the tournament in order to answer top-k discriminative typicality queries. The other instances are only used to compute the approximate discriminative typicality scores of the instances in O.

The cost of discriminative typicality computation within one group is $O(\frac{m}{m+n}t^2)$. Since there are $O(\frac{m+n}{t})$ groups in total, the complexity of selecting the final winner is $O(mt)$. If we run each tournament v times for better accuracy, then the overall complexity of answering a top-k discriminative typicality query is $O(kvtm)$.

4.3.2 Local Typicality Approximation

Similar to the local simple typicality approximation discussed in Section 4.2.1, we can define the local discriminative typicality as follows.

Definition 4.4 (Local discriminative typicality). Given two uncertain objects objects O and S on attributes $A_1, \cdots, A_n$ and a neighborhood threshold σ, let $\mathscr{U}$ and $\mathscr{V}$ be the random vectors generating O and S, respectively, the **local discriminative typicality** of an instance $o \in O$ on attributes $A_{i_1}, \cdots, A_{i_l}$ is defined as $LDT(o, \mathscr{U}, \mathscr{V}, \sigma) = LT(o, \{o\}, \mathscr{U}, \sigma) - LT(o, \{o\}, \mathscr{V}, \sigma)$, where $LT(o, \{o\}, \mathscr{U}, \sigma)$ and $LT(o, \{o\}, \mathscr{V}, \sigma)$ are the local simple typicality values of o in $\mathscr{U}$ and $\mathscr{V}$, respectively.

$LDT(o, \mathscr{U}, \mathscr{V}, \sigma)$ can be estimated using $LDT(o, O, S, \sigma) = LT(o, \{o\}, O, \sigma) - LT(o, \{o\}, S, \sigma)$, where $LT(o, \{o\}, O, \sigma)$ and $LT(o, \{o\}, S, \sigma)$ are the estimators of local simple typicality values of o in O and S, respectively.

Similar to Theorem 4.1, we have the following quality guarantee of local discriminative typicality approximation.

Theorem 4.5 (Local discriminative typicality approximation). *Given two uncertain objects O and S and a neighborhood threshold σ, let $\widetilde{o} = \arg\max_{o_1 \in O}\{LDT(o_1, O, S, \sigma)\}$ be the instance in O having the largest local discriminative typicality value, and $o = \arg\max_{o_2 \in O}\{DT(o, O, S)\}$ be the instance in O having the largest discriminative typicality value. Then,*

$$DT(o, O, S) - DT(\widetilde{o}, O, S) < \frac{2}{h\sqrt{2\pi}} e^{-\frac{\sigma^2}{2h^2}} \tag{4.8}$$

Moreover, for any $x \in O$,

$$|DT(x, O, S) - LDT(x, O, S, \sigma)| < \frac{1}{h\sqrt{2\pi}} e^{-\frac{\sigma^2}{2h^2}} \tag{4.9}$$

Proof. For any instance $x \in O$,

$$DT(x, O, S) = \frac{1}{h|O|} \sum_{y \in O} G_h(x, y) - \frac{1}{h|S|} \sum_{z \in (S)} G_h(x, z) \tag{4.10}$$

For simplicity, let us denote $LN(\{x\}, S, \sigma)$ by N. We have

$$\sum_{y \in O} G_h(x, y) = \sum_{y_1 \in O \cap N} G_h(x, y_1) + \sum_{y_2 \in (O - N)} G_h(x, y_2)$$

and

$$\sum_{z \in S} G_h(x,z) = \sum_{z_1 \in S \cap N} G_h(x,z_1) + \sum_{z_2 \in S-N)} G_h(x,z_2)$$

Because

$$LDT(x,O,S) = \frac{1}{h|O|} \sum_{y_1 \in O \cap N} G_h(x,y_1) - \frac{1}{h|(S)|} \sum_{z_1 \in S \cap N} G_h(x,z_1)$$

Equation 4.10 can be rewritten as

$$\begin{aligned} & DT(x,O,S) - LDT(x,O,S) \\ & = \tfrac{1}{h|O|} \sum_{y_2 \in (O-N)} G_h(x,y_2) - \frac{1}{h|S|} \sum_{z_2 \in (S-N)} G_h(x,z_2) \end{aligned} \tag{4.11}$$

According to the definition of local neighborhood, $d(x,y_2) > \sigma$ holds for any $y_2 \in (O-N)$, and $|O| > |O-N|$. Thus,

$$0 < \frac{1}{h|O|} \sum_{y_2 \in (O-N)} G_h(x,y_2) < \frac{1}{h\sqrt{2\pi}} e^{-\frac{\sigma^2}{2h^2}}$$

Similarly, $d(x,z_2) > \sigma$ holds for any $z_2 \in (S-N)$, and $|S| > |S-N|$. Thus,

$$0 < \frac{1}{h|S|} \sum_{z_2 \in (S-N)} G_h(x,z_2)) < \frac{1}{h\sqrt{2\pi}} e^{-\frac{\sigma^2}{2h^2}}$$

Therefore,

$$|\tfrac{1}{h|O|} \sum_{y_2 \in (O-N)} G_h(x,y_2) - \frac{1}{h|S|} \sum_{z_2 \in (S-N)} G_h(x,z_2)| < \frac{1}{\sqrt{2\pi}} e^{-\frac{\sigma^2}{2h^2}} \tag{4.12}$$

Inequality 4.9 follows from Inequality 4.11 and 4.12 immediately.

Applying Inequality 4.9 to o and $\widetilde{o}$, we have

$$|DT(o,O,S) - LDT(o,O,S)| < \frac{1}{h\sqrt{2\pi}} e^{-\frac{\sigma^2}{2h^2}}$$

and

$$|DT(\widetilde{o},O,S) - LDT(\widetilde{o},O,S)| < \frac{1}{h\sqrt{2\pi}} e^{-\frac{\sigma^2}{2h^2}}$$

Since $LDT(\widetilde{o},O,S-O) \geq LDT(o,O,S-O)$, we have

$$\begin{aligned} DT(o,O,S) - DT(\widetilde{o},O,S) = & \quad (DT(o,O,S) - LDT(o,O,S)) \\ -(DT(\widetilde{o},O,S) - LDT(\widetilde{o},O,S)) & \quad +(LDT(o,O,S) - LDT(\widetilde{o},O,S)) \\ < \tfrac{2}{h\sqrt{2\pi}} e^{-\frac{\sigma^2}{2h^2}} & \end{aligned}$$

Inequality 4.8 is proved. ■

4.3.2.1 DLTA: Direct Local Typicality Approximation

Theorem 4.5 can be directly used to approximate discriminative typicality. Given a neighborhood threshold σ, for each instance $x \in O$, we compute the σ-neighborhood of $\{x\}$ in O and S, respectively, and thus its local discriminative typicality. Searching the σ-neighborhood can also be done using a VP-tree, as described in Section 4.2.2.

The cost of computing the local discriminative typicality of an instance $x \in O$ is $O(|LN(\{x\}, O+S, \sigma)|)$. The overall cost of computing the local discriminative typicality of all instances in the target object O is $O(\sum_{x \in O} |LN(\{x\}, O+S, \sigma)|)$. As analyzed in Section 4.2.2, the σ-neighborhood of an instance may contain the whole data set in the worst case. Thus, the time complexity of the direct local typicality approximation method for discriminative typicality is $O(m(m+n))$, where m is the cardinality of the target object O, and n is the number of instances in S. However, data is often distributed as clusters in practice, and the number of instances contained in the σ-neighborhood of each instance is often small.

4.3.2.2 LT3: Local Typicality Approximation Using Tournaments

The local typicality approximation method using tournaments (LT3) method can be extended to answer top-k discriminative typicality queries, which follows the same framework as answering top-k simple typicality queries. The only difference is that, in the tournament in each node, local discriminative typicality is computed, instead of local simple typicality.

Similar to Theorem 4.3, we have the following guarantee of the quality of answering top-k discriminative typicality queries using the LT3 method.

Theorem 4.6. *Given two uncertain objects O and S, let o be the instance in O of the largest discriminative typicality and $\widetilde{o}$ be an instance computed by the LT3 method using local discriminative typicality approximation with the tournament group size t. Then,*

$$DT(o, O, S) - DT(\widetilde{o}, O, S) < \frac{2}{h\sqrt{2\pi}} e^{-\frac{\sigma^2}{2h^2}} \cdot \lceil \log_t |S+O| \rceil$$

Proof. As indicated by Inequality 4.8 in Theorem 4.5, at each level of the tournament, an error up to $\frac{2}{h\sqrt{2\pi}} e^{-\frac{\sigma^2}{2h^2}}$ in terms of discriminative typicality is introduced. For objects O and S, there are $\lceil \log_t |S+O| \rceil$ levels of tournaments. Thus, we have the inequality in the theorem. ∎

Sampling method introduced in Section 4.2.3.3 can also be used to bound the runtime of the LT3 algorithm for discriminative typicality approximation.

4.4 Answering Representative Typicality Queries

The representative typicality for an instance o in an uncertain object O with respect to an reported answer set A was defined in Definition 2.9. In this section, we first propose a straightforward approach to find the exact answer of a top-k representative typicality query. Then, we will discuss how to extend the approximation techniques proposed for simple typicality queries to efficiently answer top-k representative typicality queries.

4.4.1 An Exact Algorithm and Complexity

When the answer set A is empty, the most representatively typical instance is simply the most typical instance o_1, which can be computed using Algorithm 4.1. After o_1 is added to A, the group typicality $GT(A,O)$ is the simple typicality score $T(o_1,O)$, since all members in uncertain object O are represented by o_1.

Then, in order to compute the next instance with the maximal representative typicality score, according to Definition 2.9, we have to compute the group typicality score $G(A \cup \{o\},O)$ for each instance $o \in (O-A)$ and select the instance with the maximal score.

To compute $GT(A \cup \{o\},O)$, we first construct $N(o,A,O)$ for instance o as follows. We scan all instances in $(O-A)$. For each instance $x \in (O-A)$, suppose $x \in N(o',A,O)$ for an instance $o' \in A$, which means that o' is the instance closest to x in A. If $d(o,x) < d(o',x)$, then x is removed from $N(o',A,O)$ and is added to $N(o,A,O)$. To make the computation efficient, we maintain the minimum distance from an instance $x \in (O-A)$ to the instances in A by a 1-dimensional array. The minimum distances are updated every time after a new object is added into A.

Then, $T(o,N(o,A,O))$, the simple typicality of o in $N(o,A,O)$, is computed using Algorithm 4.1. Probability $Pr(N(o,A,O))$ is $\frac{|N(o,A,O)|}{|O|}$. For other instances $o' \in A$, since $N(o',A,O)$ may be changed, the simple typicality scores $T(o',N(o',A,O))$ and $Pr(N(o',A,O))$ are updated accordingly. Last, $GT(A \cup \{o\},O)$ can be calculated according to Definition 2.8.

The above procedure is repeated to find the next most representatively typical instance, until k instances are found.

The complexity of the exact algorithm is $O(kn^2)$ because each time after an instance is added to A, the representative typicality scores of all instances in $(S-A)$ need to be recomputed to find the next instance with the largest representative typicality score.

4.4.2 A Randomized Tournament Method

A top-k representative typicality query can be answered using the randomized tournament method.

At the beginning, the answer set A is empty, so the randomized tournament method works exactly the same as finding the most typical instance using the randomized tournament method as described in Section 4.1.3. The winner instance of the tournament is added to A.

To compute the i-th $(i > 1)$ instance with the highest approximate representative typicality score, a randomized tournament is conducted from bottom up, similar to finding the first answer in A. The only difference is that the representative typicality score of each instance in each group is computed, instead of the simple typicality score. The instance with the maximal representative typicality score in each group is the winner and is sent to the next round of tournament. The final winner is an approximation of the i-th most representatively typical instance, and is added to A.

A top-k representative typicality query can be answered by k randomized tournaments. To ensure a higher accuracy, we can run each tournament several times, and pick the winner instance with the highest representative typicality score on the whole data set.

The complexity of the randomized tournament to find the i-th instance $(i \leq k)$ with the highest representative typicality score is $O(vtn)$, where v is the number of times the tournament is run, t is the group size, and n is the size of the data set. This is because, finding the instance with the highest representative typicality score in each group takes $O(t^2)$ time, and there are $O(\frac{n}{t})$ groups in total. To answer a top-k representative typicality query, k randomized tournaments need to be conducted. Therefore, the overall complexity is $O(kvtn)$.

4.4.3 Local Typicality Approximation Methods

The locality property in simple typicality approximation can be extended to address the representative typicality approximation.

Let A be the current reported answer set. The local group typicality of A is computed by only considering the instances in the σ-neighborhood of $o \in A$. The intuition is, if an instance is not in the σ-neighborhood of o, then the contribution from o to this instance is small and can be ignored.

Definition 4.5 (Local group typicality). Given an uncertain object O, a neighborhood threshold σ and a subset of instances $A \subset O$, let $\mathscr{X}$ be the random vector that generates the samples O, the **local group typicality** of A is

$$LGT(A, \mathscr{X}, \sigma) = \sum_{o \in A} LT(o, \{o\}, \mathscr{X}_{D(o,A)}, \sigma) \cdot Pr(N)$$

where $LT(o, \{o\}, \mathscr{X}_{D(o,A)}, \sigma)$ is the local simple typicality of o in its representing region $D(o,A)$ and $N = D(o,A) \cap D(\{o\}, \sigma)$ is the σ-neighborhood region of o in its representing region $D(o,A)$. ■

Definition 4.6 (Local representative typicality). Given an uncertain object O, a neighborhood threshold σ and a **reported answer set** $A \subset O$, let $\mathscr{X}$ be the random vector generating the samples O, the **local representative typicality** of an object $o \in (O - A)$ is $LRT(o, A, \mathscr{X}, \sigma) = LGT(A \cup \{o\}, \mathscr{X}, \sigma) - LGT(A, \mathscr{X}, \sigma)$. ■

For any instance $o \in A$, let $N(o, A, O) = \{x | x \in O \cap D(o,A)\}$ be the set of instances in O that lie in $D(o,A)$, then $LN(\{o\}, N(o,A,O), \sigma)$ is the σ-neighborhood of o in $N(o,A,O)$. The local group typicality $LGT(A, \mathscr{X}, \sigma)$ can be estimated using $LGT(A, O, \sigma) = \sum_{o \in A} LT(o, \{o\}, N(o,A,O), \sigma) \cdot \frac{|LN(\{o\}, N(o,A,O), \sigma)|}{|O|}$. Hence, the local representative typicality is estimated by $LRT(o, A, O, \sigma) = LGT(A \cup \{o\}, O, \sigma) - LGT(A, O, \sigma)$.

Local representative typicality can approximate representative typicality with good quality, as shown below.

Theorem 4.7 (Local representative typicality approximation). *Given an uncertain object O, a neighborhood threshold σ, and an answer set $A \subset O$, let $\widetilde{o} = \arg\max_{o_1 \in (O-A)}\{LRT(o_1, A, O, \sigma)\}$ be the instance in $(O - A)$ having the largest local representative typicality value, and $o = \arg\max_{o_2 \in (O-A)}\{RT(o, A, O)\}$ be the instance in $(O - A)$ having the largest representative typicality value. Then,*

$$RT(o, A, O) - RT(\widetilde{o}, A, O) < \frac{2}{h\sqrt{2\pi}} e^{-\frac{\sigma^2}{2h^2}} \tag{4.13}$$

Moreover, for any $x \in (O - A)$,

$$|RT(x, A, O) - LRT(x, A, O, \sigma)| < \frac{1}{h\sqrt{2\pi}} e^{-\frac{\sigma^2}{2h^2}} \tag{4.14}$$

Proof. To prove Theorem 4.7, we need the following lemma.

Lemma 4.1 (Local group typicality score approximation). *Given an uncertain object O, a neighborhood threshold σ and a reported answer set $A \subset O$.*

$$GT(A, O) - LGT(A, O, \sigma) < \frac{1}{h\sqrt{2\pi}} e^{-\frac{\sigma^2}{2h^2}} \tag{4.15}$$

Proof. For each instance $o \in A$, let $N(o, A, O)$ be the set of instances in O that lie in $D(o,A)$, according to Equation 4.1, we have

$$T(o, N(o,O,A)) \cdot Pr(N(o,O,A)) = \frac{1}{h|N(o,O,A)|} \sum_{x \in N(o,O,A)} G_h(x,o) \times \frac{|N(o,O,A)|}{|O|}$$
$$= \frac{1}{h|O|} \sum_{x \in N(o,O,A)} G_h(x,o)$$

Let $N = LN(\{o\}, N(o,A,O), \sigma)$ be the σ-neighborhood of o in $N(o,A,O)$, then

$$LT(o,\{o\},N(o,O,A),\sigma)\cdot Pr(N) = \frac{1}{|N|}\sum_{x\in N} G_h(x,o)\times\frac{|N|}{|O|} = \frac{1}{|O|}\sum_{x\in N} G_h(x,o)$$

Thus,

$$\begin{aligned}&T(o,N(o,O,A))Pr(N(o,O,A)) - LT(o,\{o\},N(o,O,A),\sigma)Pr(N)\\ &= \tfrac{1}{h|O|}\left(\sum_{x\in N(o,O,A)} G_h(x,o) - \sum_{x\in N} G_h(x,o)\right) = \tfrac{1}{h|O|}\sum_{x\in N(o,O,A)-N} G_h(x,o)\end{aligned}$$

For instance $x \in N(o,O,A) - N$, x is not in the σ-neighborhood of o, so $d(x,o) > \sigma$. Therefore

$$\tfrac{1}{h|O|}\sum_{x\in N(o,O,A)-N} G_h(x,o) < \tfrac{1}{h|O|\sqrt{2\pi}}\sum_{x\in N(o,O,A)-N} e^{-\frac{\sigma^2}{2h^2}}$$

Thus, we have

$$\begin{aligned}>(A,O) - LGT(A,O,\sigma)\\ &= \sum_{o\in A}\left(T(o,N(o,O,A))Pr(N(o,O,A)) - LT(o,\{o\},N(o,O,A),\sigma)Pr(N)\right)\\ &= \sum_{o\in A}\tfrac{1}{h|O|}\sum_{x\in N(o,O,A)-N} G_h(x,o) < \tfrac{1}{h|O|}\sum_{o\in A}\sum_{x\in N(o,O,A)-N}\tfrac{1}{\sqrt{2\pi}}e^{-\frac{\sigma^2}{2h^2}}\\ &< \tfrac{1}{h\sqrt{2\pi}}e^{-\frac{\sigma^2}{2h^2}}\end{aligned}$$

Equation 4.15 holds.

Proof of Theorem 4.7. For any instance $x \in O$,

$$RT(o,A,O) = GT(A\cup\{o\},O) - GT(A,O)$$

$$LRT(o,A,O,\sigma) = LGT(A\cup\{o\},O,\sigma) - LGT(A,O,\sigma)$$

Therefore,

$$\begin{aligned}&RT(o,A,O) - LRT(o,A,O,\sigma)\\ &= (GT(A\cup\{o\},O) - LGT(A\cup\{o\},O,\sigma)) - (GT(A,O) - LGT(A,O,\sigma))\end{aligned}$$

Using Lemma 4.1, we have

$$0 \le GT(A\cup\{o\},O) - LGT(A\cup\{o\},O,\sigma) < \frac{1}{h\sqrt{2\pi}}e^{-\frac{\sigma^2}{2h^2}}$$

and

$$0 \le GT(A,O) - LGT(A,O,\sigma) < \frac{1}{h\sqrt{2\pi}}e^{-\frac{\sigma^2}{2h^2}}$$

Thus,

$$-\frac{1}{h\sqrt{2\pi}}e^{-\frac{\sigma^2}{2h^2}} < RT(o,A,O) - LRT(o,A,O,\sigma) < \frac{1}{h\sqrt{2\pi}}e^{-\frac{\sigma^2}{2h^2}}$$

Inequality 4.14 follows from the above inequality immediately.

Applying Inequality 4.14 to o and $\widetilde{o}$, we have

$$|RT(o,A,O) - LRT(o,A,O,\sigma)| < \frac{1}{h\sqrt{2\pi}}e^{-\frac{\sigma^2}{2h^2}}$$

and

$$|RT(\widetilde{o},A,O) - LRT(\widetilde{o},A,O,\sigma)| < \frac{1}{h\sqrt{2\pi}}e^{-\frac{\sigma^2}{2h^2}}$$

Since $LRT(\widetilde{o},A,O,\sigma) \geq LRT(o,A,O,\sigma)$

$$\begin{aligned}
&RT(o,A,O) - RT(\widetilde{o},A,O)\\
&= (RT(o,A,O) - LRT(o,A,O,\sigma)) - (RT(\widetilde{o},A,O) - LRT(\widetilde{o},A,O,\sigma))\\
&+(LRT(o,A,O,\sigma) - LRT(\widetilde{o},A,O,\sigma)) \leq \frac{2}{h\sqrt{2\pi}}e^{-\frac{\sigma^2}{2h^2}}
\end{aligned}$$

Inequality 4.13 is shown. ■

4.4.3.1 DLTA: Direct Local Typicality Approximation

Direct local representative typicality approximation (DLTA) follows the similar framework of the exact algorithm described in Section 4.4.1. The only difference is that the local representative typicality score instead of the representative typicality score is computed.

To compute the local representative typicality score of an instance o given answer set A, one critical step is to compute the local group typicality score $LGT(A \cup \{o\}, O, \sigma)$. The computation is similar to the exact algorithm elaborated in Section 4.4.1, except that the local simple typicality instead of the simple typicality of o in $N(o,A,O)$ is used to compute $LGT(A \cup \{o\}, O, \sigma)$.

Suppose the current reported answer A_i ($0 \leq i < k$, $A_0 = \emptyset$) contains the first i answers to a top-k representative typicality query, computing the local simple typicality of an instance $o \in (O - A_i)$ takes $O(|LN(\{o\}, N(o,A,O), \sigma)|)$, where $LN(\{o\}, N(o,A,O), \sigma)$ is the σ-neighborhood of o in the set of its represented members $N(o,A,O)$. Thus, the complexity of computing the $(i+1)$-th answer is $O(\sum_{o \in (O-A_i)} |LN(\{o\}, N(o,A,O), \sigma)|)$.

The overall complexity of answering a top-k representative typicality query is $O(\sum_{i=0}^{k-1} \sum_{o \in (O-A_i)} |LN(\{o\}, N(o,A,O), \sigma)|)$. In the worst case, the local neighborhood of any instance o may contain the whole data set. Moreover, A_i contains i objects, so $|O - A_i| = n - i$. Therefore, the overall complexity of the DLTA algorithm is $O(\sum_{i=0}^{k-1} ((n-i) \cdot n)) = O(n \cdot \frac{(2n-k-1)k}{2}) = O(kn^2)$.

4.4.3.2 LT3: Local Typicality Approximation Using Tournaments

The LT3 algorithm for simple typicality approximation can be used to answer top-k representative typicality queries. To find the instance with the largest representative typicality score. After the first answer instance is added into A, to find the approximation of the next most representatively typical instance, a tournament is conducted from bottom up. In each node of the LT-tree, we compute the local representative typicality of each instance, and select the instance with the greatest local representative typicality score as the winner, and let it go to the tournament in the parent node. The computation of local representative typicality is similar to the local representative typicality computation in the DLTA algorithm.

There is one critical difference between the LT3 algorithm for simple typicality computation and the LT3 algorithm for representative typicality computation. In simple typicality computation, once the first winner instance is computed, we only need to re-conduct part of the tournament to find the next winner instance. However, the representative typicality score of each instance changes once the reported answer set is updated. Thus, no results can be reused. A new tournament among the rest of the instances should be conducted from bottom up completely. Therefore, the LT3 method for representative typicality computation is not as efficient as the LT3 method for simple typicality or discriminative typicality computation.

Similar to Theorem 4.3, the quality of answering top-k representative typicality queries using the LT3 method is guaranteed.

Theorem 4.8. *In an uncertain object O, let o be the instance of the largest representative typicality and $\widetilde{o}$ be an instance computed by the LT3 method using local representative typicality approximation and the tournament group size t. Then,*

$$RT(o,A,O) - RT(\widetilde{o},A,O) < \frac{2}{h\sqrt{2\pi}} e^{-\frac{\sigma^2}{2h^2}} \cdot \lceil \log_t |S| \rceil$$

Proof. As indicated by Inequality 4.13 in Theorem 4.7, at each level of tournament, an error up to $\frac{2}{h\sqrt{2\pi}} e^{-\frac{\sigma^2}{2h^2}}$ in terms of the difference of representative typicality is introduced. For an uncertain object O, there are $\lceil \log_t |O| \rceil$ levels of tournaments. ■

4.5 Empirical Evaluation

In this section, we report a systematic empirical study using real data sets and synthetic data sets. All experiments were conducted on a PC computer with a 3.0 GHz Pentium 4 CPU, 1.0 GB main memory, and a 160 GB hard disk, running the Microsoft Windows XP Professional Edition operating system. Our algorithms were implemented in Microsoft Visual C++ V6.0.

Category	size	Most typical	Most discriminative typical	Most atypical
Mammal	40	Boar, Cheetah, Leopard, Lion, Lynx, Mongoose, Polecat, Puma, Raccoon, Wolf ($T = 0.16$)	Boar, Cheetah, Leopard, Lion, Lynx, Mongoose, Polecat, Puma, Raccoon, Wolf ($DT = 0.08$)	Platypus ($T = 0.01$)
Bird	20	Lark, Pheasant, Sparrow, Wren ($T = 0.15$)	Lark, Pheasant, Sparrow, Wren ($DT = 0.04$)	Penguin ($T = 0.04$)
Fish	14	Bass, Catfish, Chub, Herring, Piranha ($T = 0.15$)	Bass, Catfish, Herring, Chub, Piranha ($DT = 0.03$)	Carp ($T = 0.03$)
Invertebrate	10	Crayfish, Lobster ($T = 0.16$)	Crayfish, Lobster ($DT = 0.01$)	Scorpion ($T = 0.08$)
Insect	8	Moth, Housefly ($T = 0.13$)	Gnat ($DT = 0.02$)	Honeybee ($T = 0.06$)
Reptile	5	Slowworm ($T = 0.17$)	Pitviper ($DT = 0.007$)	Seasnake ($T = 0.08$)
Amphibian	3	Frog ($T = 0.2$)	Frog ($DT = 0.008$)	Newt, Toad ($T = 0.16$)

Table 4.2 The most typical, the most discriminatively typical, and the most atypical animals (T =simple typicality value, DT =discriminative typicality value).

4.5.1 Typicality Queries on Real Data Sets

In this section, we use two real data sets to illustrate the effectiveness of typicality queries on real applications.

4.5.1.1 Typicality Queries on the Zoo Data Set

We use the Zoo Database from the UCI Machine Learning Database Repository[1], which contains 100 tuples on 15 Boolean attributes and 2 numerical attributes, such as *hair* (Boolean), *feathers* (Boolean) and *number of legs* (numeric). All tuples are classified into 7 categories (*mammals*, *birds*, *reptiles*, *fish*, *amphibians*, *insects* and *invertebrates*).

We can consider each category as an uncertain object and each tuple as an instance. The Euclidean distances are computed between instances by treating Boolean values as binary values. We apply the simple typicality, discriminative typicality and representative typicality queries on the Zoo Database. The results of the three queries all match the common sense of typicality.

We compute the simple typicality for each animal in the data set. Table 4.2 shows the most typical and the most atypical animals of each category. Since some tuples, such as those 10 most typical animals in category *mammals*, have the same values on all attributes, they have the same typicality value. The most typical animals returned in each category can serve as good exemplars of the category. For example,

[1] http://www.ics.uci.edu/~mlearn/MLRepository.html.

Top-10 most representatively typical animals				Top-10 most typical animals			
Rank	Animal	Representative typicality score	Category	Rank	Animal	Simple typicality score	Category
1	Boar	0.0874661	Mammal	1	Boar	0.0855363	Mammal
2	Lark	0.135712	Bird	1	Cheetah	0.0855363	Mammal
3	Bass	0.176546	Fish	1	Leopard	0.0855363	Mammal
4	Gnat	0.198142	Insect	1	Lion	0.0855363	Mammal
5	Aardvark	0.213342	Mammal	1	Lynx	0.0855363	Mammal
6	wallaby	0.225642	Mammal	1	Mongoose	0.0855363	Mammal
7	starfish	0.236235	Invertebrate	1	Polecat	0.0855363	Mammal
8	Slug	0.246638	Invertebrate	1	Puma	0.0855363	Mammal
9	Dolphin	0.236347	Mammal	1	Raccoon	0.0855363	Mammal
10	frog	0.265012	Amphibian	1	Wolf	0.0855363	Mammal

Table 4.3 The most representatively typical and the most typical animals.

in category *mammals*, the most typical animals are more likely to be referred to as a mammal than the most atypical one, *platypuses*, which are one of the very few mammal species that lay eggs instead of giving birth to live young.

We apply *discriminative typicality analysis* on the Zoo Database to find the discriminative typical animals for each category. The results are listed in Table 4.2 as well. In some categories, the instances having the largest simple typicality value also have the highest discriminative typicality value, such as categories *mammals*, *birds*, *fish*, *invertebrates*, and *amphibians*. In some categories such as *insects* and *reptiles*, the most typical animals are not the most discriminatively typical. For example, in category *reptiles*, the most discriminatively typical animal is *pitvipers* in stead of *slowworm*, because *slowworm* is also similar to some animals in other categories besides *reptiles*, such as *newts* in category *amphibians*. On the other hand, *pitvipers* are venomous. Very few animals in the other categories are venomous. The result matches the above analysis.

In some situations, the results from the simple typicality queries may have a bias on the most popular categories. Table 4.3 lists the top-10 most typical animals in the Zoo Database. The 10 animals are all mammals, since *mammals* are the largest category in the Zoo Database. As a result, the top-10 most typical animals as a whole is not representative. The animals in other categories cannot be represented well.

Representative typicality queries avoid this problem. The top-10 most representatively typical animals are also listed in Table 4.3, which cover 6 out of the 7 categories in the Zoo Database. The only missed category is *Reptile*, which only contains 5 animals. *Boar* is the first animal in the answer set of both queries, since it is the most typical animal in the whole data set. Note that the representative typicality score and the simple typicality score of *boar* are slightly different, because the bandwidth parameter h is computed via sampling, and thus may have a small difference in each computation. The second most representatively typical animal is *Lark*, which is the most typical animal in the second most popular category *Bird*. *Dolphin* is in the answer to the top-10 representative typicality query, since it represents a set of aquatic mammal in the Zoo database, such as porpoise and mink. They are not typical mammals, but they are an important category if we would like to explore different kinds of mammals

To show the difference between typicality analysis and clustering analysis, we apply the k-medoids clustering algorithm [83] to the Zoo data set. We first compute the 10 clusters of the Zoo data set. The median animals of the clusters are {*Starfish, Boar, Lark, Tuatara, Dolphin, Flea, Bass, Mink, Scorpion, Hare*}. The group typicality score of the set of median animals is 0.182. At the same time, the group typicality score of the answer set of the top-10 most representatively typical animals shown in Table 4.3 is 0.216. Therefore, the set of animals found by the clustering analysis is only 84% as representative as the set of animals found by the top-k representative typicality queries.

4.5.1.2 Typicality Queries on the NBA Data Set

Name	T	Position	Minuts	Points per game	3 point throw	Rebounds	Assists	Blocks
D. Granger	0.0383	Forwards	22.6	7.5	1.6	4.9	1.2	0.8
D. George	0.0382	Forwards	21.7	6.3	3	3.9	1.0	0.5
M. Finley	0.0378	Guards	26.5	10.1	5	3.2	1.5	0.1

Table 4.4 The most typical NBA players in 2005-2006 Season (T for simple typicality values).

Top-10 simple typicality query				Top-10 discriminative typicality query				Top-10 representative typicality query			
Name	T	3PT	AST	Name	DT	3PT	AST	Name	RT	3PT	AST
R. Murray	0.076	2.4	2.6	D. West	0.0095	4.3	4.6	R. Murray	0.076	2.4	2.6
M. Jaric	0.075	2.3	3.9	D. Wesley	0.0092	5.2	2.9	A. Owens	0.06	0.8	0.4
K. Bogans	0.074	3.7	1.8	S. Claxton	0.0092	1.1	4.8	C. Bell	0.037	4.1	2.2
K. Martin	0.074	3.4	1.4	E. Jones	0.0085	6.8	2.4	M. James	0.023	6.9	5.8
A. Johnson	0.072	2.8	4.3	C. Duhon	0.0083	5.2	5	B. Barry	0.017	3.8	1.7
J. Rose	0.072	3.2	2.5	T.J. Ford	0.0082	1.9	6.6	C. Cheaney	0.017	0.2	0.5
M. Finley	0.071	5	1.5	J. Rose	0.0082	3.2	2.5	A. Acker	0.014	1.2	0.8
C. Atkins	0.071	5.3	2.8	K. Hinrich	0.0079	5.8	6.4	L. Barbosa	0.009	4.9	2.8
C. Duhon	0.071	5.2	5	J. Terry	0.0078	7.3	3.8	Z. Planinic	0.007	1.2	1
E. Jones	0.07	6.8	2.4	M. Miller	0.0077	6.5	2.7	K. Dooling	0.007	1.2	2.2

Table 4.5 The answers to top-10 simple typicality/discriminative typicality/representative typicality queries on the NBA guards (T for simple typicality values, DT for discriminative typicality values, RT representative typicality values, and 3PT for 3 point throw).

Category	Median/mean/most typical	Name	Simple typicality	# games	Min. per game
All players	median	R. Gomes	0.0271	61	22.6
	mean	N/A	0.0307	54.4	20.51
	most typical	D. Granger	0.3830	78	22.6
Centers	median	J. Voskuhl	0.0903	51	16
	mean	N/A	0.0679	52.42	17.36
	most typical	F. Elson	0.1041	72	21.9
Forwards	median	A. Jefferson	0.0747	59	18
	mean	N/A	0.0509	54.83	19.97
	most typical	M. Taylor	0.0910	67	18.1
Guards	median	C. Bell	0.0488	59	21.7
	mean	N/A	0.0230	54.54	21.73
	most typical	R. Murray	0.0756	76	27.8

Table 4.6 Comparison among medians, means, and typical players in the NBA data set.

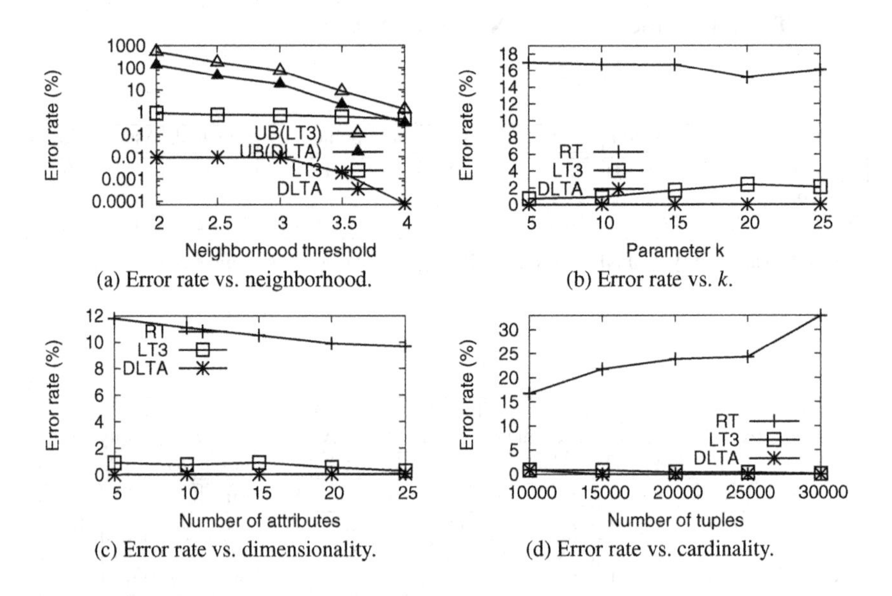

(a) Error rate vs. neighborhood.

(b) Error rate vs. *k*.

(c) Error rate vs. dimensionality.

(d) Error rate vs. cardinality.

Fig. 4.4 Approximation quality of answering top-*k* simple typicality queries.

We apply typicality queries on the NBA 2005-2006 Season Statistics[2]. The data set contains the technical statistics of 458 NBA players, including 221 guards, 182 forwards and 55 centers, on 16 numerical attributes.

As discussed in Section 2.2.1, we can model the set of NBA players as an uncertain object and each player as an instance of the object. Table 4.4 shows the top-3 most typical players, and some of the attribute values. The results answer Jeff's question in Section 2.2.1.

To answer Jeff's question in Section 2.2.1, we model the set of guards as the target uncertain object, and the set of forwards and centers as the other two uncer-

[2] http://sports.yahoo.com/nba/stats/.

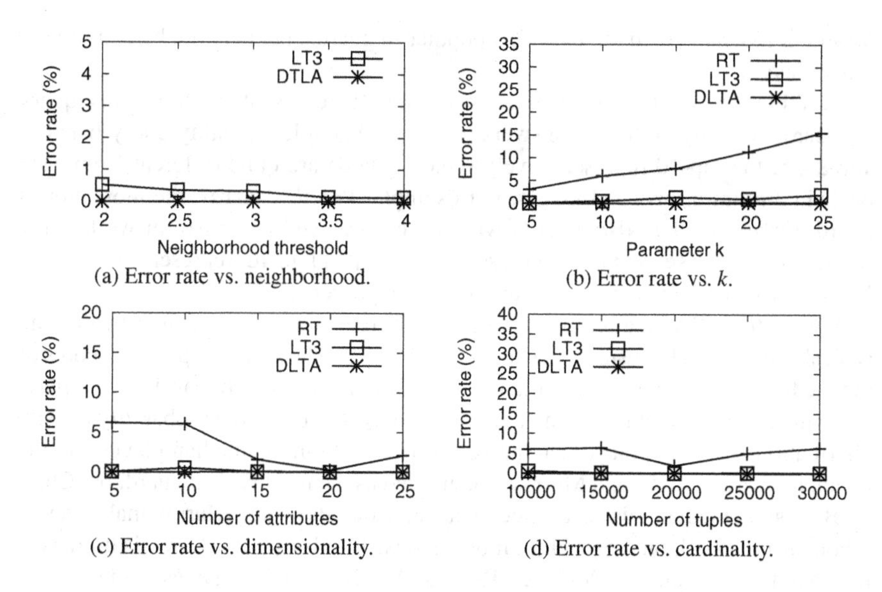

(a) Error rate vs. neighborhood.

(b) Error rate vs. k.

(c) Error rate vs. dimensionality.

(d) Error rate vs. cardinality.

Fig. 4.5 Approximation quality of answering top-k discriminative typicality queries.

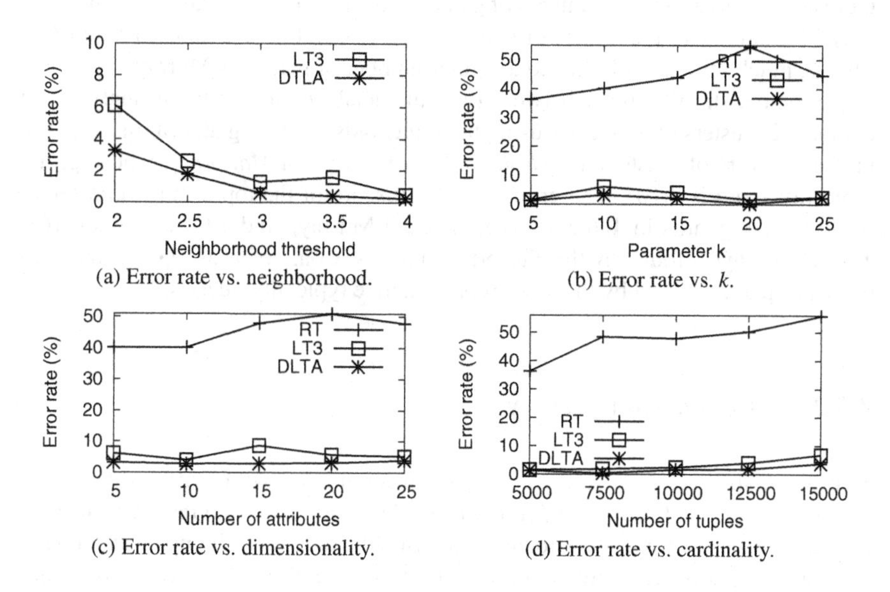

(a) Error rate vs. neighborhood.

(b) Error rate vs. k.

(c) Error rate vs. dimensionality.

(d) Error rate vs. cardinality.

Fig. 4.6 Approximation quality of answering top-k representative typicality queries.

tain objects in the data set. We conduct a top-10 discriminative typicality query on guards. The results are shown in Table 4.5. For comparison, in the same table we also list the answer to the top-10 simple typicality query on guards. To explain the results, we list some selected attributes as well. The most discriminatively typical guards have better performance than those of the highest simple typicality in *3 point throws* or *assists*, which are the skills popular in guards, but may not be common in other players.

In Table 4.5, we list the answers to the top-10 representative typicality query on guards. Comparing to the answers to a top-10 simple typicality query listed in Table 4.5, the top-10 representatively typical guards are quite different from each other in 3 point throws and assists. For example, *Ronald Murray*, the most typical guard, represents the NBA guards who are experienced and perform well, while *Andre Owens*, the second most representatively typical guard, represents a group of NBA guards whose performances are relatively poorer.

We use the NBA data set to examine the differences among medians, means, and typical instances. The results are shown in Table 4.6. The simple typicality scores of the medians and the means are often substantially lower than the most typical players, which justifies that the geometric centers may not reflect the probability density distribution. A typical player can be very different from the median player and the mean. For example, Ronald Murray is identified as the most typical guard, but Charlie Bell is the median guard. The technical statistics show that Murray makes fewer rebounds than Bell, but contributes more assists. To this extent, Murray is more typical than Bell as a guard. Moreover, Ronald Murray played 76 games in the season, while Charlie Bell only played 59 games. If we take the range $76 \pm 6 = [70, 82]$, then there are 92 guards whose numbers of games played are in the range; while there are only 31 guards whose numbers of games played are in the range $59 \pm 6 = [53, 65]$. That is, much more guards played a similar number of games as Murray.

To compare the difference between typicality analysis and clustering analysis, we compute 2 clusters of all guards using the k-medoids clustering algorithm [83]. The median players of clusters are {*Ronald Murray, Stephon Marbury*}, whose group typicality score is 0.105. The group typicality score of the top-2 most representatively typical guards in Table 4.5 (i.e.,{Ronald Murray, Andre Owens}) is 0.161. The set of players found by the clustering analysis is only 65% as representative as the set of players found by the top-k representative typicality queries.

4.5.2 Approximation Quality

To evaluate the query answering quality on uncertain data with large number of instances, we use the Quadraped Animal Data Generator also from the UCI Machine Learning Database Repository to generate synthetic data sets with up to 25 numeric attributes. We test the approximation quality of the RT (randomized tournament) method, the DLTA (direct local typicality approximation) method, and the LT3 (local typicality approximation using tournaments) method on top-k simple typicality queries, top-k discriminative typicality queries, and top-k representative typicality queries, respectively. The results are reported in the rest of this section.

First of all, we test RT, DLTA, and LT3 for top-k simple typicality queries. To measure the error made by an approximation algorithm, we use the following *error rate* measure. For a top-k typicality query Q, let A be the set of k instances returned

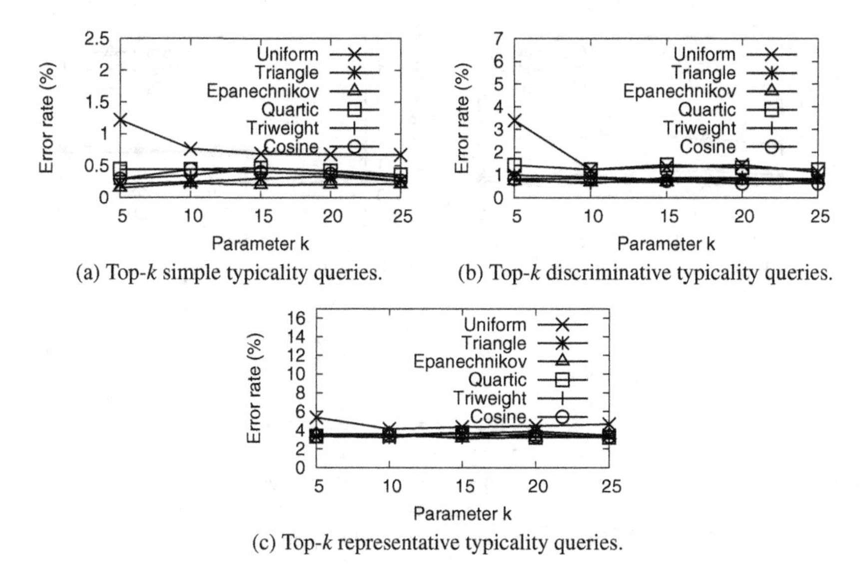

Fig. 4.7 The error rates of using different kernel functions with respect to k.

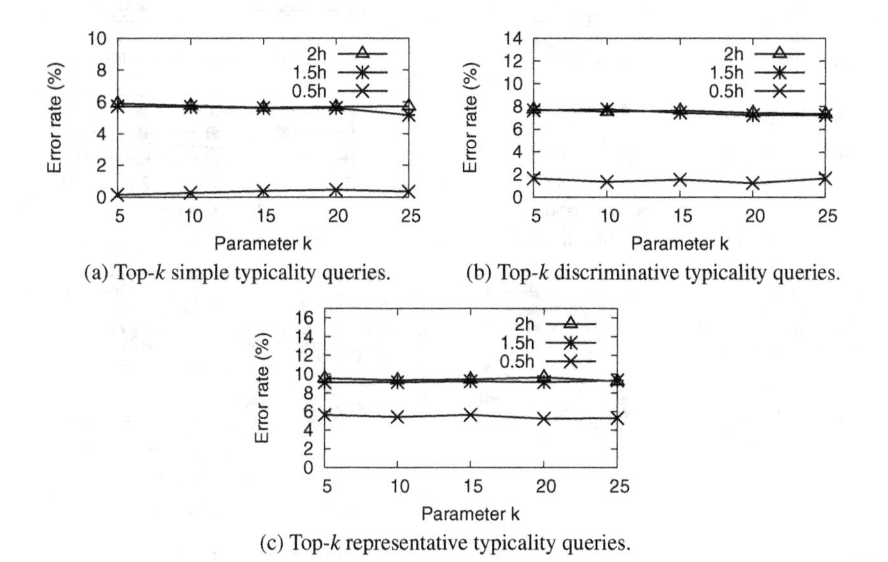

Fig. 4.8 The error rates of using different bandwidth values with respect to k.

by the exact algorithm, and $\widetilde{A}$ be the set of k instances returned by an approximation algorithm. Then, the error rate e is

$$e = \frac{\sum_{o \in A} T(o,O) - \sum_{o \in \widetilde{A}} T(o,O)}{\sum_{o \in A} T(o,O)} \times 100\% \tag{4.16}$$

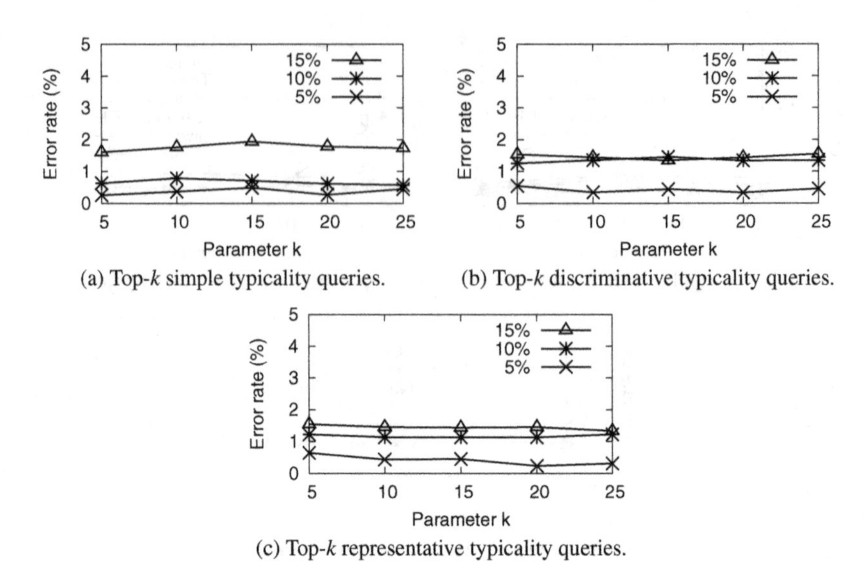

Fig. 4.9 The error rates of having different amount of noises with respect to k.

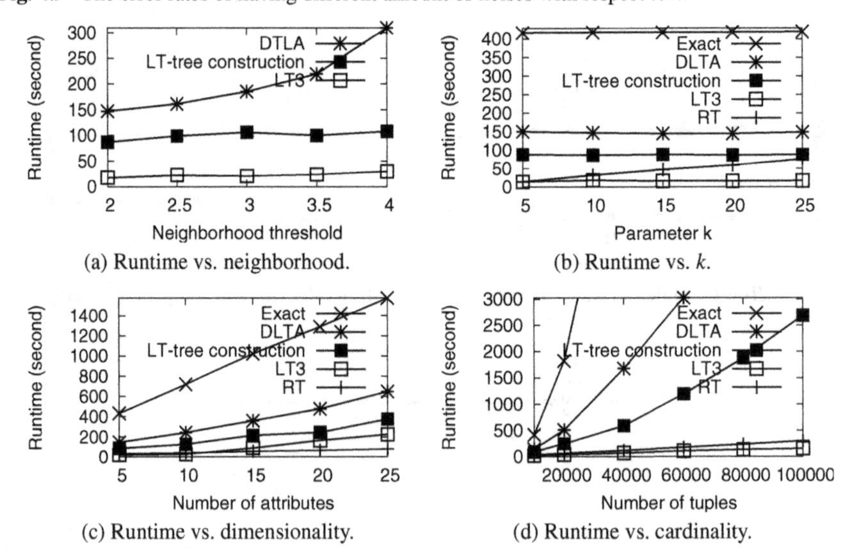

Fig. 4.10 Efficiency and scalability of answering top-k simple typicality queries.

The error rate of the exact algorithm is always 0.

We compare three approximation techniques: the randomized tournament method (RT, Algorithm 4.2), the direct local typicality approximation method (DLTA, Section 4.2.2), and the LT-tree tournament method (LT3, Section 4.2.3). By default, we set the number of instances to 10,000, the dimensionality to 5 attributes, and conduct top-10 simple typicality queries. When local typicality is computed, by default

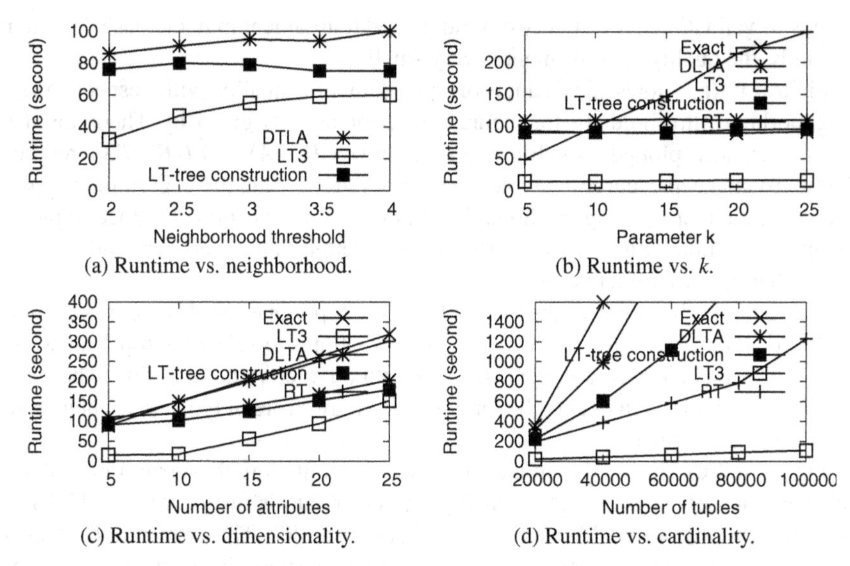

(a) Runtime vs. neighborhood. (b) Runtime vs. k.

(c) Runtime vs. dimensionality. (d) Runtime vs. cardinality.

Fig. 4.11 Efficiency and scalability of answering top-k discriminative typicality queries.

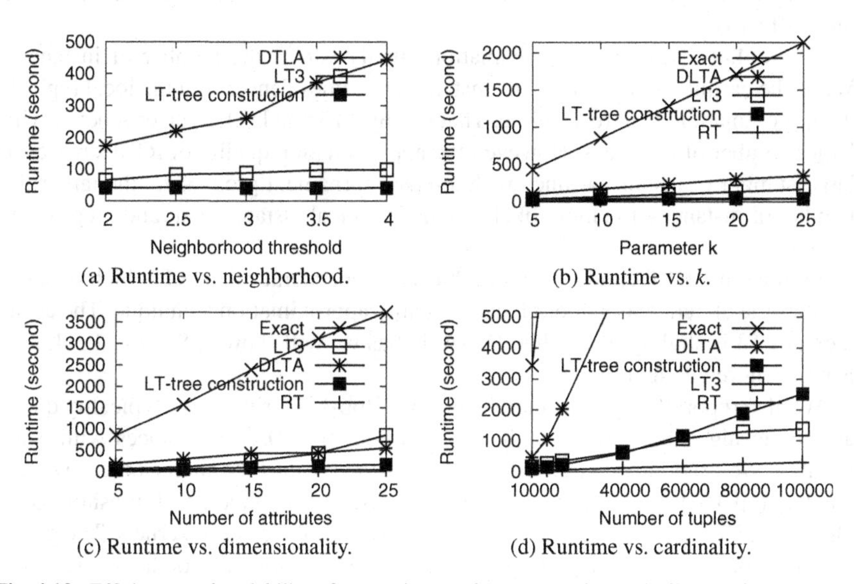

(a) Runtime vs. neighborhood. (b) Runtime vs. k.

(c) Runtime vs. dimensionality. (d) Runtime vs. cardinality.

Fig. 4.12 Efficiency and scalability of answering top-k representative typicality queries.

we set the neighborhood threshold to $2h$, where h is the bandwidth of the Gaussian kernel. In such a case, according to Theorem 4.1, the difference between the simple typicality value and the local simple typicality value of any instance is always less than 0.05. In the randomized tournament method, by default the tournament group size is 10 and 4 times validation are conducted. We observe that although with more rounds of validations, the quality of randomized tournament may increase, but after 3 rounds, the quality improvement is very small.

Figure 4.4(a) shows the change of approximation quality with respect to the neighborhood threshold. In the figure, the error bounds given by Theorems 4.1 and 4.3 are also plotted, which are labeled as $UB(DLTA)$ and $UB(LT3)$, respectively. To make the curves legible, the error rates are in the logarithmic scale. Clearly, the larger the neighborhood threshold, the more accurate the local typicality approximation. Our methods perform much better than the error bounds, which shows that they are effective in practice.

In Figure 4.4(b), we vary the value of k in the top-k queries. The approximation quality of RT is not sensitive to k, since it runs k times to select the top-k answers. Both DLTA and LT3 see a larger error rate with a larger value of k, this is because the distant neighbors may get a better chance to play a role in typicality when more instances are returned.

Figure 4.4(c) shows the impact of dimensionality on the error rate. DLTA achieves the best approximation quality, the error rate is up to 0.066%. LT3 has an accuracy close to DLTA, and is much better than RT. The error rate decreases as the dimensionality increases, since the average pairwise distance among the instances in the object also increases and the local typicality approximation becomes more effective.

Figure 4.4(d) tests the approximation quality versus the number of instances. When the cardinality increases, the instances becomes denser, and the local typicality approximation is more accurate. That is why LT3 and DLTA perform better with larger number of instances. However, the approximation quality of RT decreases in large number of instances, since with a fixed tournament group size, the larger the number of instances, the more likely the most typical instance in a random group is biased.

In summary, DLTA and LT3 both achieve better accuracy than RT, which strongly justifies the effectiveness of our local typicality approximation technique. The accuracy of LT3 is slightly lower than DLTA, but as we will show in Section 4.5.4, LT3 is much more efficient than DLTA.

We also report the approximation quality of top-k discriminative typicality query answering algorithms. By default, the data contains 10,000 instances with 5 attributes. We randomly assign 20% of the instances into the target object O, and conduct top-10 discriminative typicality queries. The neighborhood threshold σ of DLTA and LT3 is set to $2h$, where h is the bandwidth of Gaussian kernels. The group size of randomized tournament is set to 50, and 4 validations are conducted. Here we increase the tournament size to 50, since only 20% instances are actually involved in the tournament, as we explained in Section 4.3.1.

The error rate measure is defined as follows. We normalize $DT(o,O,S-O)$ as $DT'(o,C,S-C) = DT(o,O,S-O) - \min_{x \in C} DT(x,O,S-O)$, in order to make the $DT(o,O,S-O)$ value always non-negative. For a top-k discriminative typicality query Q, let A be the set of k instances returned by the exact algorithm, and $\widetilde{A}$ be the set of k instances returned by an approximation algorithm. Then, the error rate e is

$$e = \frac{\sum_{o \in A} DT'(o,O,S-O) - \sum_{o \in \widetilde{A}} DT'(o,O,S-O)}{\sum_{o \in A} DT'(o,O,S-O)} \cdot 100\% \quad (4.17)$$

The approximation quality of the three methods are shown in Figure 4.5. In general, the comparison among RT, DLTA and LT3 is similar to the situation of the top-k simple typicality query evaluation listed in Figure 4.4.

To test the approximation quality for representative typicality, we conducted various experiments. By default, the data set contains $5,000$ instances with 5 attributes, and conduct top-10 representative typicality queries. The neighborhood threshold σ of DLTA and LT3 is set to $2h$, where h is the bandwidth of Gaussian kernels. The group size of randomized tournament is set to 10, and 4 validations are conducted.

We adopt the following error rate measure. For a top-k representative typicality query Q, let A be the set of k instances returned by the exact algorithm, and $\widetilde{A}$ be the set of k instances returned by an approximation algorithm. $GT(A,O)$ and $GT(\widetilde{A},O)$ are the group typicality scores of A and $\widetilde{A}$, respectively. Then, the error rate e is

$$e = \frac{|GT(A,O) - GT(\widetilde{A},O)|}{GT(A,O)} \times 100\% \quad (4.18)$$

The error rate measure computes the difference between the group typicality of the exact answer and the group typicality of the approximate answer. If the error rate is small, even the instances in the two answer sets are different, the approximation to the answer still represents the whole data set well. The approximation quality of representative typicality approximation is shown in Figure 4.6. The explanations are similar to the situations of simple typicality queries.

In summary, for all three types of typicality queries, DLTA has the best approximation quality, while RT gives the largest error rates. LT3 has comparable approximation quality to DLTA.

4.5.3 Sensitivity to Parameters and Noise

To test the sensitivity of the answers of top-k typicality queries with respect to the kernel function and the bandwidth value, we use the Quadraped Animal Data Generator from the UCI Machine Learning Database Repository. to generate synthetic data sets with $10,000$ instances and 5 attributes.

We first fix the bandwidth value $h = \frac{1.06s}{\sqrt[5]{n}}$ as discussed in Section 4.1.1, and use the kernel functions listed in Table 4.1 to answer top-k simple typicality/ discriminative typicality/ representative typicality queries. We compare the results computed using the Gaussian kernel function and the results computed using some other kernel functions as follows.

Let the results returned by using the Gaussian kernel be A and the results returned by using other kernel functions be $\widetilde{A}$, the error rates of the answers to the three typicality queries are computed using Equations 4.16, 4.17 and 4.18, respectively. The curves are shown in Figure 4.7. The results match the discussion in Section 4.1.1:

the answers computed using some other kernel functions are similar to the answers computed using the Gaussian kernel, since the error rates are very low.

We then use the Gaussian kernel function and vary the bandwidth value from $0.5h$ to $2h$, where $h = \frac{1.06s}{\sqrt[5]{n}}$ is the default bandwidth value used in other experiments. Let A be the answer set computed using the default bandwidth value h and $\widetilde{A}$ be the answer set computed using other bandwidth values, the error rates are computed using Equations 4.16, 4.17 and 4.18, respectively. From the results shown in Figure 4.8, we can see that the answers computed using different bandwidth values are similar in their typicality/discriminative typicality/group typicality score. Moreover, using smaller bandwidth values causes less difference than using larger bandwidth values. Larger bandwidth values smooth out the peaks of the density curves, which are the most typical points.

In summary, the answers to top-k simple typicality queries, top-k discriminative typicality queries and top-k representative typicality queries are insensitive to the choice of kernel functions and the bandwidth values.

Moreover, we evaluate the sensitivity of top-k typicality queries with respect to noise in data sets. We use the Quadraped Animal Data Generator to generate synthetic data sets with $10,000$ instances and 5 attributes. In addition, we add 5% to 15% noise instances whose attribute values are uniformly distributed in the domain of each attribute. Gaussian kernel function and bandwidth $h = \frac{1.06s}{\sqrt[5]{n}}$ are used. The answers returned are denoted by $\widetilde{A}$, and the answers computed when removing the noises are denoted by A. The error rates in Figure 4.9(a), 4.9(b), and 4.9(c) are computed using Equations 4.16, 4.17 and 4.18, respectively. Clearly, the results of top-k typicality queries are not sensitive to noise and outliers in data sets.

4.5.4 Efficiency and Scalability

To test the efficiency and the scalability of our methods, we report in Figures 4.10, 4.11, and 4.12 the runtime in the experiments conducted in Figures 4.4, 4.5, and 4.6, respectively.

As shown in Figure 4.10(a), the runtime of DLTA increases substantially when the neighborhood threshold increases, but the increase of runtime for the LT3 method is mild, thanks to the tournament mechanism.

Figure 4.10(b) shows that the runtime of DLTA and LT3 is insensitive to the increase of k. LT3 incrementally computes other top-k answers after the top-1 answer is computed. Thus, computing more answers only takes minor cost. The RT method has to run the tournaments k rounds, and thus the cost is linear to k.

As shown in Figure 4.10(c), among the four methods, RT is the fastest and the exact algorithm is the slowest. LT3 and DLTA are in between, and LT3 is faster than DLTA. All methods are linearly scalable with respect to dimensionality.

Figure 4.10(d) shows the scalability of the four algorithms with respect to database size. RT has a linear scalability. LT3 clearly has the better performance and scalability than DLTA on large data sets.

The trends for discriminative typicality queries and representative typicality queries are similar, as shown in Figure 4.11 and Figure 4.12, respectively.

In summary, as RT has linear complexity, when runtime is the only concern, RT should be used. While DLTA and LT3 are much more scalable than the exact algorithm and are more accurate than RT, they are good when both accuracy and efficiency matter. LT3 has the better efficiency and scalability than DLTA, while achieving comparable accuracy to DLTA.

4.6 Summary

In this chapter, we discussed how to compute the answers to top-k typicality queries defined in Section 2.2.1.

- We applied kernel density estimation to compute the likelihood of instances in an uncertain object and presented an exact query answering algorithm. We showed that the complexity of answering top-k typicality queries is quadratic.
- We developed a linear-time randomized algorithm which adopts a tournament mechanism and computes the approximation answers to top-k typicality queries.
- The randomized algorithm does not provide a quality guarantee in the approximated answers, therefore, we further explored the locality nature of kernel density estimation and proposed two approximation algorithms that can provide good quality guarantees.

By a systematic empirical evaluation using both real data sets and synthetic data sets, we illustrated the effectiveness of top-k typicality queries, and verified the accuracy and the efficiency of our methods.

Chapter 5
Probabilistic Ranking Queries on Uncertain Data

In this chapter, we discuss how to answer probabilistic ranking queries defined in Section 2.2.2 on the probabilistic database model. The techniques developed in this chapter will also be used in Chapter 6.

Given a probabilistic table T with a set of generation rules $\mathscr{R}$ and a top-k selection query $Q^k_{P,f}$, where P is a predicate, f is a scoring function, and k is a positive integer, the rank-k probability of a tuple $t \in T$ is the probability that t is ranked at the k-th position in possible worlds according to $Q^k_{P,f}$, that is

$$Pr(t,k) = \sum_{W \in \mathscr{W} \text{ s.t. } t = W_f(k)} Pr(W)$$

where $W_f(k)$ denotes the tuple ranked at the k-th position in W.

Moreover, the top-k probability of t is the probability that t is ranked top-k in possible worlds according to f, that is,

$$Pr^k(t) = \sum_{j=1}^{k} Pr(t,j)$$

Last, the *p-rank* of t is the minimum k such that $Pr^k(t) \geq p$, denoted by

$$MR_p(t) = \min\{k | Pr^k(t) \geq p\}$$

Four types of probabilistic ranking queries are developed.

- A **probabilistic threshold top-k query**(*PT-k query* for short) [30, 31] finds the tuples whose top-k probabilities are at least a probability threshold p $(0 < p \leq 1)$.
- A **rank threshold query** (*RT-k query* for short) retrieves the tuples whose p-ranks are at most k. RT-k queries are reverse queries of PT-k queries.
- A **top-(k,l) query** [32, 33] finds the top-l tuples with the highest top-k probabilities $(l > 0)$.
- A **top-(p,l) query** returns the top-l tuples with the smallest p-ranks. Top-(p,l) queries are reverse queries of top-(k,l) queries.

A naïve method of answering probabilistic ranking queries is to enumerate all possible worlds and apply the query to each possible world. Then, we can compute the top-k probability and p-rank of each tuple and select the tuples satisfying the queries. Unfortunately, the naïve method is inefficient since, as discussed before, there can be a huge number of possible worlds on an uncertain table. In [49], Dalvi and Suciu showed that even the problem of counting all possible worlds with distinct top-k lists is #P-Complete. Therefore, enumerating all possible worlds is too costly on large uncertain data sets. That motivates our development of efficient algorithms which avoid searching all possible worlds.

In this chapter, we first discuss the efficient top-k probability computation in Section 5.1 and present an efficient exact algorithm in Section 5.2. Then, we develop a fast sampling algorithm and a Poisson approximation based algorithm in Sections 5.3 and 5.4, respectively. Last, to support efficient online query answering, we propose *PRist+*, a compact index, in Section 5.5. An efficient index construction algorithm and efficient query answering methods are developed for *PRist+*. An empirical study in Section 5.6 using real and synthetic data sets verifies the effectiveness of probabilistic ranking queries and the efficiency of our methods.

5.1 Top-k Probability Computation

In this section, we first introduce how to compute the exact rank-k probability values. Top-k probabilities and p-ranks can be directly derived from rank-k probabilities.

5.1.1 The Dominant Set Property

Hereafter, by default we consider a top-k selection query $Q^k_{P,f}$ on an uncertain table T. $P(T) = \{t | t \in T \wedge P(t) = true\}$ is the set of tuples satisfying the query predicate. $P(T)$ is also an uncertain table where each tuple in $P(T)$ carries the same membership probability as in T. Moreover, a generation rule R in T is projected to $P(T)$ by removing all tuples from R that are not in $P(T)$.

$P(T)$ contains all tuples satisfying the query, as well as the membership probabilities and the generation rules. Removing tuples not in $P(T)$ does not affect the rank-k probabilities of the tuples in $P(T)$. Therefore, we only need to consider $P(T)$ in computing rank-k probabilities. To keep our discuss simple, we use T to denote the set of tuples satisfying query predicate P.

For a tuple $t \in T$ and a possible world W such that $t \in W$, whether $t \in W_f(k)$ depends only on how many other tuples in T ranked higher than t appear in W.

Definition 5.1 (Dominant set). Given a scoring function f on a probabilistic table T, for a tuple $t \in T$, the **dominant set** of t is the subset of tuples in T that are ranked higher than t, i.e., $S_t = \{t' | t' \in T \wedge t' \prec_f t\}$. ■

Theorem 5.1 (The dominant set property). *For a tuple $t \in T$, $Pr^k_{Q,T}(t) = Pr^k_{Q,S_t}(t)$, where $Pr^k_{Q,T}(t)$ and $Pr^k_{Q,S_t}(t)$ are the top-k probabilities of t computed using tuples in T and in S_t, respectively.*

Proof. The theorem follows with the definition of top-k probability directly. ∎

Using the dominant set property, we scan the tuples in T in the ranking order and derive the rank-k probabilities for each tuple $t \in T$ based on the tuples preceding t. Generation rules involving multiple tuples are handled by the rule-tuple compression technique developed later in this section.

5.1.2 The Basic Case: Independent Tuples

We start with the basic case where all tuples are independent. Let $L = t_1 \cdots t_n$ be the list of all tuples in table T in the ranking order. Then, in a possible world W, a tuple $t_i \in W$ $(1 \leq i \leq n)$ is ranked at the j-th $(j > 0)$ position if and only if exactly $(j-1)$ tuples in the dominant set $S_{t_i} = \{t_1, \ldots, t_{i-1}\}$ also appear in W. The *subset probability* $Pr(S_{t_i}, j)$ is the probability that j tuples in S_{t_i} appear in possible worlds.

Trivially, we have $Pr(\emptyset, 0) = 1$ and $Pr(\emptyset, j) = 0$ for $0 < j \leq n$. Then, the rank-k probability can be computed as

$$Pr(t_i, k) = Pr(t_i)Pr(S_{t_i}, k-1)$$

Moreover, the top-k probability of t_i is given by

$$Pr^k(t_i) = \sum_{j=1}^{k} Pr(t_i, j) = Pr(t_i) \sum_{j=1}^{k} Pr(S_{t_i}, j-1)$$

Particularly, when $i \leq k$, we have $Pr^k(t_i) = Pr(t_i)$.

The following theorem can be used to compute the subset probability values efficiently.

Theorem 5.2 (Poisson binomial recurrence). *In the basic case, for $1 \leq i, j \leq |T|$,*

1. $Pr(S_{t_i}, 0) = Pr(S_{t_{i-1}}, 0)(1 - Pr(t_i)) = \prod_{j=1}^{i}(1 - Pr(t_i))$;
2. $Pr(S_{t_i}, j) = Pr(S_{t_{i-1}}, j-1)Pr(t_i) + Pr(S_{t_{i-1}}, j)(1 - Pr(t_i))$.

Proof. In the basic case, all tuples are independent. The theorem follows with the basic probability principles. This theorem is also called the Poisson binomial recurrence in [103]. ∎

5.1.3 Handling Generation Rules

In general, a probabilistic table may contain some multi-tuple generation rules. For a tuple $t \in T$, two situations due to the presence of multi-tuple generation rules complicate the computation.

First, there may be a rule R such that some tuples involved in R are ranked higher than t. Second, t itself may be involved in a generation rule R. In both cases, some tuples in S_t are dependent and thus Theorem 5.2 cannot be applied directly. Can dependent tuples in S_t be transformed to independent ones so that Theorem 5.2 can still be used?

Let $T = t_1 \cdots t_n$ be in the ranking order, i.e., $t_i \preceq_f t_j$ for $i < j$. We compute $Pr^k(t_i)$ for a tuple $t_i \in T$. A multi-tuple generation rule $R : t_{r_1}, \cdots, t_{r_m}$ $(1 \leq r_1 < \cdots < r_m \leq n)$ can be handled in one of the following cases.

Case 1: $t_i \preceq_f t_{r_1}$, i.e., t_i is ranked higher than or equal to all tuples in R. According to Theorem 5.1, R can be ignored.

Case 2: $t_{r_m} \prec_f t_i$, i.e., t_i is ranked lower than all tuples in R. We call R *completed* with respect to t_i.

Case 3: $t_{r_1} \prec_f t_i \preceq_f t_{r_m}$, i.e., t_i is ranked in between tuples in R. R is called *open* with respect to t_i. Among the tuples in R ranked better than t_i, let $t_{r_{m_0}} \in R$ be the lowest ranked tuple i.e., $r_{m_0} = \max_{l=1}^{m}\{r_l < i\}$. The tuples involved in R can be divided into two parts: $R_{left} = \{t_{r_1}, \ldots, t_{r_{m_0}}\}$ and $R_{right} = \{t_{r_{m_0}+1}, \ldots, t_{r_m}\}$. $Pr^k(t_i)$ is affected by tuples in R_{left} only and not by those in R_{right}. Two subcases may arise, according to whether t belongs in R or not: in subcase 1, $t_i \notin R$; in subcase 2, $t_i \in R$, i.e., $t_i = t_{r_{m_0}+1}$.

Since in Case 1, generation rule R can be ignored, in the rest of this section, we mainly discuss how to handle generation rules in Case 2 and Case 3.

We first consider computing $Pr^k(t_i)$ when an generation rule $R : t_{r_1} \oplus \cdots \oplus t_{r_m}$ $(1 \leq r_1 < \cdots < r_m \leq n)$ is involved.

In Case 2, t_i is ranked lower than all tuples in R. At most one tuple in R can appear in a possible world. According to Theorem 5.1, we can combine all tuples in R into an *generation rule-tuple* t_R with membership probability $Pr(R)$.

Corollary 5.1 (Generation rule-tuple compression). *For a tuple $t \in T$ and a multi-tuple generation rule R, if $\forall t' \in R$, $t' \prec_f t$, then $Pr^k_{Q,T}(t) = Pr^k_{Q,T(R)}(t)$ where $T(R) = (T - \{t|t \in R\}) \cup \{t_R\}$, tuple t_R takes any value such that $t_R \prec_f t$, $Pr(t_R) = Pr(R)$, and other generation rules in T remain the same in $T(R)$.* ∎

In Case 3, t_i is ranked between the tuples in R, which can be further divided into two subcases. First, if $t_i \notin R$, similar to Case 2, we can compress all tuples in R_{left} into an generation rule-tuple $t_{r_1,\ldots,r_{m_0}}$ where membership probability $Pr(t_{r_1,\ldots,r_{m_0}}) = \sum_{j=1}^{m_0} Pr(t_{r_j})$, and compute $Pr^k(t_i)$ using Corollary 5.1.

Second, if $t_i \in R$, in a possible world where t_i appears, any tuples in R cannot appear. Thus, to determine $Pr^k(t_i)$, according to Theorem 5.1, we only need to consider the tuples ranked higher than t_i and not in R, i.e., $S_{t_i} - \{t'|t' \in R\}$.

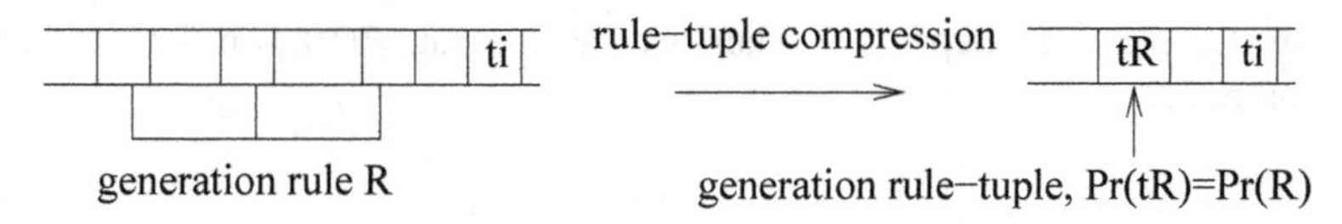

Case 2: ti is ranked lower than all tuples in R

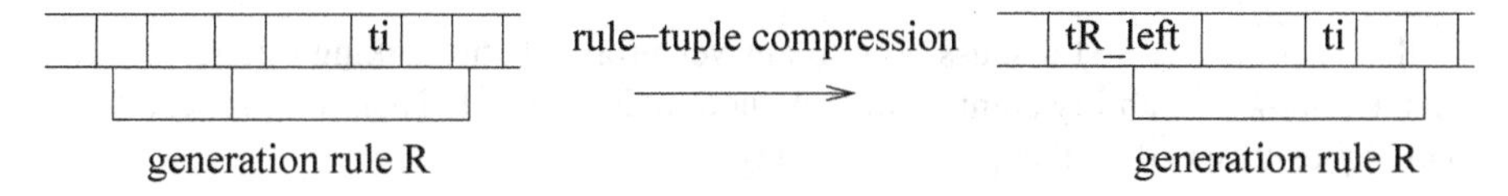

Case 3 (Subcase 1): ti is ranked between tuples in R and ti is not in R

Fig. 5.1 Computing $Pr^k(t_i)$ for one tuple t_i.

Corollary 5.2 (Tuple in generation rule). *For a tuple $t \in R$ such that $|R| > 1$, $Pr^k_{Q,T}(t) = Pr^k_{Q,T'}(t)$ where uncertain table $T' = (S_{t_i} - \{t'|t' \in R\}) \cup \{t\}$.*
Proof. Since the tuples in R are mutually exclusive, the probability that one tuple in R appears is the sum of the membership probability of each tuple in R. Therefore, the corollary holds. ∎

For a tuple t and its dominant set S_t, we can check t against the multi-tuple generation rules one by one. Each multi-tuple generation rule can be handled by one of the above two cases as illustrated in Figure 5.1, and the dependent tuples in S_t can be either compressed into some generation rule-tuples or removed due to the involvement in the same generation rule as t. After the generation rule-tuple compression, the resulting set is called the *compressed dominant set* of t, denoted by $T(t)$. Based on the above discussion, for a tuple $t \in T$, all tuples in $T(t) \cup \{t\}$ are independent, $Pr^k_{Q,T}(t) = Pr^k_{Q,T(t)\cup\{t\}}(t)$. We can apply Theorem 5.2 to calculate $Pr^k(t)$ by scanning $T(t)$ once.

Example 5.1 (Generation rule-tuple compression). Consider a list of tuples $t_1, \cdots, t_{11}$ in the ranking order. Suppose we have two multi-tuple generation rules: $R_1 = t_2 \oplus t_4 \oplus t_9$ and $R_2 = t_5 \oplus t_7$. Let us consider how to compute $Pr^3(t_6)$ and $Pr^3(t_7)$.

Tuple t_6 is ranked between tuples in R_1, but $t_6 \notin R_1$. The first subcase of Case 3 should be applied. Thus, we compress $R_{1left} = \{t_2, t_4\}$ into an generation rule-tuple $t_{2,4}$ with membership probability $Pr(t_{2,4}) = Pr(t_2) + Pr(t_4)$. Similarly, t_6 is also ranked between tuples in R_2 and $t_6 \notin R_2$, but $R_{2left} = \{t_5\}$. The compression does not remove any tuple. After the compression, $T(t_6) = \{t_1, t_{2,4}, t_3, t_5\}$. Since the tuples in $T(t_6) \cup \{t_6\}$ are independent, we can apply Theorem 5.2 to compute $Pr^3(t_6)$ using $T(t_6)$.

Since $t_7 \in R_2$, the tuples in R_2 except for t_7 itself should be removed. Thus, we have $T(t_7) = \{t_1, t_{2,4}, t_3, t_6\}$. ∎

We can sort all tuples in T into a sorted list L in the ranking order. For each tuple t_i, by one scan of the tuples in L before t_i, we obtain the compressed dominant set

$T(t_i)$ where all tuples are independent. Then, we can compute $Pr^k(t_i)$ on $T(t_i) \cup \{t_i\}$ using Theorem 5.2.

5.2 Exact Query Answering Methods

In this section, we first discuss how to answer probabilistic ranking queries based on the rank-k probability computation method in Section 5.1. Then, we propose two techniques to speed up the query answering methods.

5.2.1 Query Answering Framework

Straightforwardly, to answer a PT-k query with probability threshold p, we simply scan all tuples in T in the ranking order and compute the top-k probability of each tuple. The tuples whose top-k probabilities passing the threshold p are returned as the answers.

It is shown in Corollary 2.3 that, for a PT-k query and an RT-k query with the same rank parameter k and probability threshold p values, the answers are identical. Therefore, the above method can be directly used to answer an RT-k query.

The method can be extended to answer top-(k,l) queries as following. Again, we scan the tuples in the ranking order. A buffer B that contains at most l tuples is maintained during the scan. At the beginning, the first l tuples $t_1, \cdots, t_l$ are added into the buffer. The probability threshold p is set to the minimal probability value of $t_1, \cdots, t_l$. That is, $p = \min_{1 \le i \le l}\{Pr^k(t_i)\}$. Then, for each tuple t_i $(l+1 \le i \le |T|)$, if $Pr^k(t_i) \ge p$, then the tuple in B with the minimal top-k probability is replaced by t_i. The probability threshold p is again set to the minimal top-k probability $\min_{t \in B}\{Pr^k(t)\}$ of the tuples in the buffer. At the end of the scan, all tuples in the buffer are returned.

To answer a top-(p,l) query, we can use the similar procedure. The only difference is that a buffer B that contains the tuples with the smallest p-ranks during the scan is maintained. The rank threshold k is set to the largest p-rank of all tuples in the buffer. That is, $k = \max_{t \in B}\{MR_p(t)\}$.

From the above discussion, it is clear that all four types of queries discussed in this work can be answered based on top-k probability calculation. However, can we further improve the efficiency of the query answering methods? In Section 5.2.2, we discuss how to reuse subset probability calculation during computing the top-k probability values for all tuples. In Section 5.2.3, we develop several effective pruning techniques to improve the efficiency.

5.2.2 Scan Reduction by Prefix Sharing

We scan the dominant set S_{t_i} of each tuple $t_i \in T$ once and computes the subset probabilities $Pr(S_{t_i}, j)$. Can we reduce the number of scans of the sorted tuples to improve the efficiency of query evaluation?

To compute $Pr^k(t_i)$ using subset probability $Pr(S_{t_i}, j)$, the order of tuples in S_{t_i} does not matter. This gives us the flexibility to order tuples in compressed dominant sets of different tuples so that the prefixes and the corresponding subset probability values can be shared as much as possible. In this section, we introduce two reordering methods to achieve good sharing.

5.2.2.1 Aggressive Reordering

Consider the list $L = t_1 \cdots t_n$ of all tuples in T and a tuple t_i in L. Two observations help the reordering.

First, for a tuple t that is independent or is a rule-tuple of a completed rule with respect to t_i (Case 2 in Section 5.1.3), t is in $T(t')$ for any tuple $t' \succ_f t_i$. Thus, t should be ordered before any rule-tuple of a rule open with respect to t_i (Case 3 in Section 5.1.3).

Second, there can be multiple rules open with respect to t_i. Each such a rule R_j has a rule-tuple $t_{R_{j_{left}}}$, which will be combined with the next tuple $t' \in R_j$ to update the rule-tuple. Thus, if t' is close to t_i, $t_{R_{j_{left}}}$ should be ordered close to the rear so that the rule-tuple compression affects the shared prefix as little as possible. In other words, those rule-tuples of rules open with respect to t_i should be ordered in their next tuple indices descending order.

An *aggressive reordering method* to reorder the tuples is to always put all independent tuples and rule-tuples of completed rules before rule-tuples of open rules, and order rule-tuples of open rules according to their next tuples in the rules.

We scan all tuples in T in the ranking order. Two buffer lists, $L_{complete}$ and L_{open}, are used to help aggressive reordering. $L_{complete}$ contains all independent tuples or completed rule-tuples, while L_{open} contains all open rule-tuples during the scan. Both $L_{complete}$ and L_{open} are initialized to $\emptyset$ before the scan.

When scanning tuple t_i, we compute the compressed dominant set of t_i, and update $L_{complete}$ and L_{open} according the following two cases.

Case 1: If t_i is an independent tuple, then the compressed dominant set of t_i contains all tuples in $L_{complete}$ and L_{open}. Moreover, we put t_i into $L_{complete}$, meaning that t_i will appear in the compressed dominant set of all tuples ranked lower than t_i.

Case 2: If t_i is involved in a multi-tuple generation rule $R : t_{r_1}, \ldots, t_{r_m}$, then the compressed dominant set of t_i contains all tuples in $L_{complete}$ and L_{open}, except for the rule-tuple $t_{R_{left}}$ in L_{open}, where $t_{R_{left}}$ is the rule-tuple compressed from all tuples in R that are ranked higher than t_i.

In order to update $L_{complete}$ and L_{open}, the following two subcases arise. First, if t_i is not the last tuple in R (i.e., $t_i = t_{r_{m_0}}$ where $1 \leq m_0 < m$), then we update rule-tuple $t_{R_{left}}$ by compressing t_i into $t_{R_{left}}$, using the methods discussed in Section 5.1.3. If $t_{R_{left}}$ is not in L_{open}, then we add $t_{R_{left}}$ into L_{open}. Moreover, we sort the rule-tuples in L_{open} in their next tuple indices descending order. Second, if t_i is the last tuple in R, which means that the rule-tuple t_R will never be updated later. Therefore, we remove $t_{R_{left}}$ from L_{open}, and add t_R into $L_{complete}$.

The subset probabilities of the tuples in $L_{complete}$ only need to be calculated once and can be reused by all tuples ranked lower than them. In contrast, the rule-tuples in L_{open} may be updated when other tuples in the same rule are scanned. Therefore, only part of the subset probabilities can be reused.

For two consecutive tuples t_i and t_{i+1} in the sorted list L of all tuples in T, let $L(t_i)$ and $L(t_{i+1})$ be the sorted lists of the tuples in $T(t_i)$ and $T(t_{i+1})$, respectively, given by the aggressive reordering method. Let $Prefix(L(t_i), L(t_{i+1}))$ be the longest common prefix between $L(t_i)$ and $L(t_{i+1})$. The total number of subset probability values needed to be calculated is $Cost = \sum_{i=1}^{n-1}(|L(t_{i+1})| - |Prefix(L(t_i), L(t_{i+1}))|)$.

tuple	Aggressive reordering		Lazy reordering	
	Prefix	Cost	Prefix	Cost
t_1	$\emptyset$	0	$\emptyset$	0
t_2	$\emptyset$	0	$\emptyset$	0
t_3	$t_{1,2}$	1	$t_{1,2}$	1
t_4	$t_3 t_{1,2}$	2	$t_{1,2} t_3$	1
t_5	$t_3 t_{1,2}$	0	$t_{1,2} t_3$	0
t_6	$t_3 t_{4,5} t_{1,2}$	2	$t_{1,2} t_3 t_{4,5}$	1
t_7	$t_3 t_6 t_{4,5} t_{1,2}$	3	$t_{1,2} t_3 t_{4,5} t_6$	1
t_8	$t_3 t_6 t_7 t_{4,5}$	2	$t_3 t_6 t_7 t_{4,5}$	4
t_9	$t_3 t_6 t_7 t_{1,2,8} t_{4,5}$	2	$t_3 t_6 t_7 t_{4,5} t_{1,2,8}$	1
t_{10}	$t_3 t_6 t_7 t_9 t_{1,2,8}$	2	$t_3 t_6 t_7 t_9 t_{1,2,8}$	2
t_{11}	$t_3 t_6 t_7 t_9 t_{4,5,10}$	1	$t_3 t_6 t_7 t_9 t_{4,5,10}$	1
	Total cost: 15		Total cost: 12	

Table 5.1 Results of reordering techniques.

Example 5.2 (Aggressive reordering). Consider a list of ranked tuples $t_1, \cdots, t_{11}$ with two multi-tuple rules $R_1 : t_1 \oplus t_2 \oplus t_8 \oplus t_{11}$ and $R_2 : t_4 \oplus t_5 \oplus t_{10}$. The compressed dominant sets of tuples in the orders made by the aggressive reordering method is listed in Table 5.1.

For example, before scanning t_6, $L_{complete}$ contains independent tuple t_3 and L_{open} contains rule-tuples $t_{4,5}$ and $t_{1,2}$. $t_{4,5}$ is ranked before $t_{1,2}$, since the next tuple in R_2, t_{10}, is ranked lower than R_1's next tuple t_8. Since t_6 is independent, the compressed dominant set of t_6 includes all 3 tuples in $L_{complete}$ and L_{open}. $T(t_6)$ and $T(t_5)$ only share the common prefix t_3, therefore, the cost of calculating the subset probabilities for $T(t_6)$ is $3 - 1 = 2$. After scanning t_6, t_6 is added into $L_{complete}$.

Algorithm 5.1 The aggressive reordering algorithm

Input: an uncertain table T, a set of generation rules $\mathscr{R}$, and a top-k query Q_f^k
Output: Reordered compressed dominant set $T(t_i)$ of each tuple $t_i \in T$
Method:

```
1: set L_complete = ∅, L_open = ∅
2: retrieve tuples in T in the ranking order one by one
3: for all t_i ∈ T do
4:     if t_i is independent then
5:         T(t_{i+1}) = L_complete + L_open
6:         add t_i to the rear of L_complete
7:     else
8:         T(t_{i+1}) = L_complete + (L_open − t_{R_left})
           {*t_i is involved in rule R}
9:         if t_i is the last tuple in a rule R then
10:            remove t_{R_left} from L_open
11:            form rule-tuple t_R and add t_R to L_complete
12:        else
13:            update rule-tuple t_{R_left} in L_open by compressing t_i ∈ R
14:            sort all rule-tuples in L_open in their next tuple indices descending order.
15:        end if
16:    end if
17: end for
```

The total cost by using the aggressive reordering method is $Cost_{aggressive} = 15$. As a comparison, without reordering, the total number of subset probability values needed to be calculated is the sum of lengths of all compressed dominant sets, which is 31. ■

The aggressive reordering algorithm is given in Algorithm 5.1. The complexity of the aggressive reordering algorithm lies in the sorting of L_{open}. When scanning a tuple, the rule-tuples in L_{open} are already sorted and at most one rule-tuple in L_{open} is updated. Therefore, the complexity is $O(\log |L_{open}|)$ for processing one tuple. The overall complexity is $O(n \log n)$ where n is the number of tuples in T.

5.2.2.2 Lazy Reordering

On the other hand, a *lazy method* always reuses the longest common prefix in the compressed dominant set of the last tuple scanned, and reorders only the tuples not in the common prefix using the two observations discussed in Section 5.2.2.1.

We scan the tuples in T in the ranking order. During the scan, we maintain the compressed dominant set of the last scanned tuple. When processing tuple t_i, one of the following two cases may apply.

Case 1: If t_i is an independent tuple, or t_i is the first tuple scanned in a multi-tuple generation rule R, then the compressed dominant set of t_i can be computed by one of the following two subcases.

First, if t_{i-1} is independent, then $T(t_i)$ can be obtained by adding t_{i-1} to the rear of $T(t_{i-1})$.

Second, if t_{i-1} is involved in a multi-tuple generation rule R', then $T(t_i)$ is computed by adding $t_{R'_{left}}$ to the rear of $T(t_{i-1})$.

Case 2: If t_i is involved in a multi-tuple generation rule R but not the first tuple scanned in R, then there are three subcases.

First, if t_{i-1} is involved in the same rule with t_i, then $T(t_i) = T(t_{i-1})$.

Second, if t_{i-1} is an independent tuple, then $T(t_{i-1})$ must contain a rule-tuple $t_{R_{left}}$ corresponding to rule R, which should not be included in $T(t_i)$. Moreover, t_{i-1} should be added at the rear of $T(t_{i-1})$. In that case, the longest common prefix of $T(t_{i-1})$ and $T(t_i)$ includes the tuples ranked before $t_{R_{left}}$ in $T(t_{i-1})$. The subset probabilities for the tuples in the longest common prefix can be reused. For those tuples or rule-tuples not in the longest common prefix, we reorder them so that the independent tuples are always sorted before the rule-tuples and the rule-tuples are sorted in their next tuple indices descending order.

Third, if t_{i-1} is involved in another rule $R' \neq R$, then there are two differences between $T(t_{i-1})$ and $T(t_i)$: 1) $T(t_{i-1})$ contains $t_{R_{left}}$ but $T(t_i)$ does not; 2) $T(t_i)$ includes $t_{R'_{left}}$ but $T(t_{i-1})$ does not. Therefore, we first add $t_{R'_{left}}$ to the rear of $t_{R'_{left}}$. Then, as discussed in the second subcase, we can reuse the longest common prefix of $T(t_{i-1})$ and $T(t_i)$, and reorder the tuples not in the longest common prefix.

Example 5.3 (Lazy reordering). Consider a list of ranked tuples $t_1, \cdots, t_{11}$ with multi-tuple rules $R_1 : t_1 \oplus t_2 \oplus t_8 \oplus t_{11}$ and $R_2 : t_4 \oplus t_5 \oplus t_{10}$ again. The compressed dominant sets of tuples in the orders made by the lazy reordering method is listed in Table 5.1.

The lazy reordering method orders the compressed dominant sets in the same way as the aggressive reordering method for t_1, t_2 and t_3.

For t_4, the aggressive method orders t_3, an independent tuple, before $t_{1,2}$, the rule-tuple for rule R_1 which is open with respect to t_4. The subset probability values computed in $T(t_3)$ cannot be reused. The lazy method reuses the prefix $t_{1,2}$ from $T(t_3)$, and appends t_3 after $t_{1,2}$. All subset probability values computed in $T(t_3)$ can be reused. The total cost of the lazy reordering method is 12. ■

The algorithm of the lazy reordering method is given in Algorithm 5.2. The complexity is $O(n \log n)$, as analyzed in the aggressive reordering method.

We can show that the lazy method is never worse than the aggressive method.

Theorem 5.3 (Effectiveness of lazy reordering). *Given a ranked list of tuples in T, let $Cost(agg)$ and $Cost(lazy)$ be the total number of subset probability values needed to be calculated by the aggressive reordering method and the lazy reordering method, respectively. Then, $Cost(agg) \geq Cost(lazy)$.*

Proof. For two consecutive tuples t_i and t_{i+1} in T $(1 \leq i \leq |T| - 1)$, we consider the following three cases.

Algorithm 5.2 The lazy reordering algorithm

Input: an uncertain table T, a set of generation rules $\mathscr{R}$, and a top-k query Q^k_f
Output: Reordered compressed dominant set $T(t_i)$ of each tuple $t_i \in T$
Method:

```
1:  retrieve tuples in T in the ranking order one by one
2:  set T(t_1) = ∅
3:  for all t_i ∈ T (i ≥ 2) do
4:     if t_i is independent or t_i is the first tuple in R then
5:        if t_{i-1} is independent then
6:           T(t_i) = T(t_{i-1}) + t_{i-1}
7:        else
8:           T(t_i) = T(t_{i-1}) + t_{R'_left}
           {*t_{i-1} ∈ R'}
9:        end if
10:    else
11:       if t_{i-1} is involved in R then
12:          T(i) = T(i-1)
13:       else
14:          if t_{i-1} is involved in rule R' then
15:             T(t_i) = T(t_{i-1}) - t_{R_left} + t_{R'_left}
16:          else
17:             T(t_i) = T(t_{i-1}) - t_{R_left} + t_{i-1}
18:          end if
19:          reorder all tuples in T(t_i) that are ranked lower than t_{R_left} in their next tuple indices
             descending order.
20:       end if
21:    end if
22: end for
```

First, if t_i and t_{i+1} are involved in the same generation rule, then the cost of computing $T(t_{i+1})$ is 0 using either aggressive reordering or lazy reordering, since $T(t_{i+1})$ contains the same set of tuples in $T(t_i)$.

Second, if t_{i+1} is an independent tuple or the first tuple in a generation rule R, then the cost of computing $T(t_{i+1})$ using lazy reordering is 1, since a tuple t_i (if t_i is independent) or rule-tuple $t_{R'_{left}}$ (if t_i is involved in R') should be added into $T(t_i)$ to form $T(t_{i+1})$. The cost of computing $T(t_{i+1})$ using aggressive reordering is at least 1.

Third, if t_{i+1} is involved in rule R but is not the first tuple in R, then $t_{R_{left}}$ must be removed from $T(t_i)$. Moreover, t_i or $t_{R'_{left}}$ should be added into $T(t_i)$, as discussed in the second case. Let $L_{reorder}$ be the set of tuples or rule-tuples in $T(t_i)$ that are ranked lower than $t_{R_{left}}$, then the cost of computing $T(t_{i+1})$ is $|L_{reorder}| + 1$. Now let us show that, using aggressive reordering, the same amount of cost is also needed before scanning t_{i+1}. For each tuple $t \in L_{reorder}$, one of the following two subcases may arise: 1) t is an independent tuple or a completed rule-tuple, then t must be put into $L_{complete}$ using aggressive reordering. The subset probability of t_R need to be recomputed once $L_{complete}$ is updated. Thus, 1 cost is required. 2) t is an open-rule tuple, then it must be put into L_{open} using aggressive reordering. The subset probability of t needs to be recomputed after removing L_R, which requires a cost of 1.

Therefore, in any of the three cases, the cost of lazy reordering is not more than the cost of aggressive reordering. The conclusion folows. ∎

5.2.3 Pruning Techniques

So far, we implicitly have a requirement: all tuples in T are scanned in the ranking order. However, a probabilistic ranking query or reverse query is interested in only those tuples passing the query requirement. Can we avoid retrieving or checking all tuples satisfying the query predicates?

Some existing methods such as the well known TA algorithm [4] can retrieve in batch tuples satisfying the predicate in the ranking order. Using such a method, we can retrieve tuples in T progressively in the ranking order. Now, the problem becomes how we can use the tuples seen so far to prune some tuples ranked lower in the ranking order.

Consider rank parameter k and probability threshold p. We give four pruning rules: Theorems 5.4 and 5.5 can avoid checking some tuples that cannot satisfy the probability threshold, and Theorems 5.6 and 5.7 specify stopping conditions. The tuple retrieval method (e.g., an adaption of the TA algorithm [4]) uses the pruning rules in the retrieval. Once it can determine all remaining tuples in T fail the probability threshold, the retrieval can stop.

Please note that we still have to retrieve a tuple t failing the probability threshold if some tuples ranked lower than t may satisfy the threshold, since t may be in the compressed dominant sets of those promising tuples.

Theorem 5.4 (Pruning by membership probability). *For a tuple $t \in T$, $Pr^k(t) \leq Pr(t)$. Moreover, if t is an independent tuple and $Pr^k(t) < p$, then*

1. *for any independent tuple t' such that $t \preceq_f t'$ and $Pr(t') \leq Pr(t)$, $Pr^k(t') < p$; and*
2. *for any multi-tuple rule R such that t is ranked higher than all tuples in R and $Pr(R) \leq Pr(t)$, $Pr^k(t'') < p$ for any $t'' \in R$.* ∎

To use Theorem 5.4, we maintain the largest membership probability p_{member} of all independent tuples and rule-tuples for completed rules checked so far whose top-k probability values fail the probability threshold. All tuples identified by the above pruning rule should be marked failed.

A tuple involved in a multi-tuple rule may be pruned using the other tuples in the same rule.

Theorem 5.5 (Pruning by tuples in the same rule). *For tuples t and t' in the same multi-tuple rule R, if $t \preceq_f t'$, $Pr(t) \geq Pr(t')$, and $Pr^k(t) < p$, then $Pr^k(t') < p$.* ∎

Based on the above pruning rule, for each rule R open with respect to the current tuple, we maintain the largest membership probability of the tuples seen so far in R

whose top-k probability values fail the threshold. Any tuples in R that have not been seen should be tested against this largest membership probability.

Our last pruning rule is based on the observation that the sum of the top-k probability values of all tuples is exactly k. That is $\sum_{t \in T} Pr^k(t) = k$.

Theorem 5.6 (Pruning by total top-k probability). *Let A be a set of tuples whose top-k probability values have been computed. If $\sum_{t \in A} Pr^k(t) > k - p$, then for every tuple $t' \notin A$, $Pr^k(t') < p$.* ■

Moreover, we have a tight stopping condition as follows.

Theorem 5.7 (A tight stopping condition). *Let $t_1, \ldots, t_m, \ldots, t_n$ be the tuples in the ranking order. Assume $L = t_1, \ldots, t_m$ are read. Let LR be the set of open rules with respect to t_{m+1}. For any tuple t_i $(i > m)$,*

1. *if t_i is not in any rule in LR, the top-k probability of t_i $Pr^k(t_i) \leq \sum_{j=0}^{k-1} Pr(L, j)$;*
2. *if t_i is in a rule in LR, the top-k probability of t_i $Pr^k(t_i) \leq \max_{R \in LR}(1 - Pr(t_{R_{left}})) \sum_{j=0}^{k-1} Pr(L - t_{R_{left}}, j)$.*

Proof. For item (1), consider the compressed dominant set $T(t_i)$ of t_i. $L \subseteq T(t_i)$. Therefore,

$$Pr^k(t_i) = Pr(t_i) \sum_{j=0}^{k-1} Pr(T(t_i), j) \leq \sum_{j=0}^{k-1} Pr(L, j).$$

The equality holds if tuple t_{m+1} is independent with membership probability 1.

For item (2), suppose t_i is involved in an open rule $R \in LR$. $Pr(t_i) \leq 1 - Pr(t_{R_{left}})$. Moreover, for the compressed dominant set $T(t_i)$ of t_i, $(L - t_{R_{left}}) \subseteq T(t_i)$. Therefore,

$$Pr^k(t_i) = Pr(t_i) \sum_{j=0}^{k-1} Pr(T(t_i), j) \leq (1 - Pr(t_{R_{left}})) \sum_{j=0}^{k-1} Pr(L - t_{R_{left}}, j)$$

The equality holds when tuple t_{m+1} is involved in rule R' with membership probability $1 - Pr(t_{R'_{left}})$, where

$$R' = \arg\max_{R \in LR}(1 - Pr(t_{R_{left}})) \sum_{j=0}^{k-1} Pr(L - t_{R_{left}}, j).$$

■

Theorem 5.7 provides two upper bounds for tuples that have not been seen yet. If the upper bounds are both lower than the probability threshold p, then the unseen tuples do not need to be checked. The two bounds are both tight: Conclusion 1 can be achieved if $Pr(t_1) = 1$, while Conclusion 2 can be achieved if $t_i \in \arg\min_{R \in LR} Pr(t_{R_{left}})$ and $Pr(t_i) = \max_{R \in LR}\{1 - Pr(t_{R_{left}})\}$.

Algorithm 5.3 The exact algorithm with reordering and pruning techniques

Input: an uncertain table T, a set of generation rules $\mathscr{R}$, a top-k query Q^k_f, and a probability threshold p
Output: $Answer(Q,p,T)$
Method:
1: retrieve tuples in T in the ranking order one by one
2: **for all** $t_i \in T$ **do**
3: compute $T(t_i)$ by rule-tuple compression and reordering
4: compute subset probability values and $Pr^k(t_i)$
5: **if** $Pr^k(t_i) \geq p$ **then**
6: output t_i
7: **end if**
8: check whether t_i can be used to prune future tuples
9: **if** all remaining tuples in T fail the probability threshold **then**
10: exit
11: **end if**
12: **end for**

In summary, the exact algorithm for PT-k query answering is shown in Algorithm 5.3. We analyze the complexity of the algorithm as follows.

For a multi-tuple rule $R: t_{r_1}, \cdots, t_{r_m}$ where $t_{r_1}, \ldots, t_{r_m}$ are in the ranking order, let $span(R) = r_m - r_1$. When tuple t_{r_l} $(1 < l \leq m)$ is processed, we need to remove rule-tuple $t_{r_1,\ldots,r_{l-1}}$, and compute the subset probability values of the updated compressed dominant sets. When the next tuple not involved in R is processed, $t_{r_1,\ldots,r_{l-1}}$ and t_{r_l} are combined. Thus, in the worst case, each multi-tuple rule causes the computation of $O(2k \cdot span(R))$ subset probability values. Moreover, in the worst case where each tuple T passes the probability threshold, all tuples in T have to be read at least once. The time complexity of the whole algorithm is $O(kn + k\sum_{R \in \mathscr{R}} span(R))$.

As indicated by our experimental results, in practice the four pruning rules are effective. Often, only a very small portion of the tuples in T are retrieved and checked before the exact answer to a PT-k query is obtained.

Interestingly, since PT-k query answering methods can be extended to evaluate top-(k,l) queries and top-(p,l) queries (as discussed in Section 5.2.1), the pruning techniques introduced in this section can be applied to answering top-(k,l) queries and top-(p,l) queries as well.

5.3 A Sampling Method

One may trade off the accuracy of answers against the efficiency. In this section, we present a simple yet effective sampling method for estimating top-k probabilities of tuples.

For a tuple t, let X_t be a random variable as an indicator to the event that t is ranked top-k in possible worlds. $X_t = 1$ if t is ranked in the top-k list, and $X_t = 0$ otherwise. Apparently, the top-k probability of t is the expectation of X_t, i.e., $\Pr^k(t) = E[X_t]$. Our objective is to draw a set of samples S of possible worlds, and compute the mean of X_t in S, namely $E_S[X_t]$, as the approximation of $E[X_t]$.

We use uniform sampling with replacement. For table $T = \{t_1, \ldots, t_n\}$ and the set of generation rules $\mathscr{R}$, a sample unit (i.e., an observation) is a possible world. We generate the sample units under the distribution of T: to pick a sample unit s, we scan T once. An independent tuple t_i is included in s with probability $Pr(t_i)$. For a multi-tuple generation rule $R: t_{r_1} \oplus \cdots \oplus t_{r_m}$, s takes a probability of $Pr(R)$ to include one tuple involved in R. If s takes a tuple in R, then tuple t_{r_l} $(1 \leq l \leq m)$ is chosen with probability $\frac{Pr(t_{r_l})}{Pr(R)}$. s can contain at most 1 tuple from any generation rule.

Once a sample unit s is generated, we compute the top-k tuples in s. For each tuple t in the top-k list, $X_t = 1$. The indicators for other tuples are set to 0.

The above sample generation process can be repeated so that a sample S is obtained. Then, $E_S[X_t]$ can be used to approximate $E[X_t]$. When the sample size is large enough, the approximation quality can be guaranteed following from the well known Chernoff-Hoeffding bound [182].

Theorem 5.8 (Sample size). *For any $\delta \in (0,1)$, $\varepsilon > 0$, and a sample S of possible worlds, if $|S| \geq \frac{3 \ln \frac{2}{\delta}}{\varepsilon^2}$, then for any tuple t,* $\Pr\{|E_S[X_t] - E[X_t]| > \varepsilon E[X_t]\} \leq \delta$. ■

We can implement the sampling method efficiently using the following two techniques, as verified by our experiments.

First, we can sort all tuples in T in the ranking order into a sorted list L. The first k tuples in a sample unit are the top-k answers in the unit. Thus, when generating a sample unit, instead of scanning the whole table T, we only need to scan L from the beginning and generate the tuples in the sample as described before. However, once the sample unit has k tuples, the generation of this unit can stop. In this way, we reduce the cost of generating sample units without losing the quality of the sample. For example, when all tuples are independent, if the average membership probability is μ, the expected number of tuples we need to scan to generate a sample unit is $\lceil \frac{k}{\mu} \rceil$, which can be much smaller than $|T|$ when $k \ll |T|$.

Second, in practice, the actual approximation quality may converge well before the sample size reaches the bound given in Theorem 5.8. Thus, *progressive sampling* can be adopted: we generate sample units one by one and compute the estimated top-k probability of tuples after each unit is drawn. For given parameters $d > 0$ and $\phi > 0$, the sampling process stops if in the last d sample units the change of the estimated X_t for any tuple t is smaller than ϕ.

To answer a PT-k query with probability threshold p, we first run the above sampling algorithm. Then, we scan the tuples in L and output the tuples whose estimated top-k probabilities are at least p.

After obtaining estimated top-k probabilities of tuples using the above sampling method, top-(k,l) queries and top-(p,l) queries can be answered similarly as discussed in Section 5.2.1.

5.4 A Poisson Approximation Based Method

In this section, we further analyze the properties of top-k probability from the statistics aspect, and derive a general stopping condition for query answering algorithms which depends on parameter k and threshold p only and is independent from data set size. We also devise an approximation method based on the Poisson approximation. Since the PT-k query answering methods can be extended to evaluate top-(k,l) queries and top-(p,l) queries as discussed in Section 5.2.1, the Poisson approximation based method can be used to answer top-(k,l) queries and top-(p,l) queries too. We omit the details to avoid redundance.

5.4.1 Distribution of Top-k Probabilities

Let $X_1,\ldots,X_n$ be a set of independent random variables, such that $Pr(X_i = 1) = p_i$ and $Pr(X_i = 0) = 1 - p_i$ $(1 \leq i \leq n)$. Let $X = \sum_{i=1}^{n} X_i$. Then, $E[X] = \sum_{i=1}^{n} p_i$. If all p_i's are identical, $X_1,\ldots,X_n$ are called *Bernoulli trials*, and X follows a *binomial distribution*; otherwise, $X_1,\ldots,X_n$ are called *Poisson trials*, and X follows a *Poisson binomial distribution*.

For a tuple $t \in T$, the top-k probability of t is $Pr^k(t) = Pr(t)\sum_{j=1}^{k} Pr(T(t), j-1)$, where $Pr(t)$ is the membership probability of t, $T(t)$ is the compressed dominant set of t. Moreover, the probability that fewer than k tuples appear in $T(t)$ is $\sum_{j=1}^{k} Pr(T(t), j-1)$.

If there is any tuple or generation rule-tuple in $T(t)$ with probability 1, we can remove the tuple from $T(t)$, and compute the top-$(k-1)$ probability of t. Thus, we can assume that the membership probability of any tuple or rule-tuple in $T(t)$ is smaller than 1.

To compute $Pr^k(t)$, we construct a set of Poisson trials corresponding to $T(t)$ as follows. For each independent tuple $t' \in T(t)$, we construct a random trial $X_{t'}$ whose success probability $Pr(X_{t'} = 1) = Pr(t')$. For each multi-tuple generation rule $R_\oplus$ ($R_\oplus \cap T(t) \neq \emptyset$), we combine the tuples in $R_\oplus \cap T(t)$ into a rule-tuple $t_{R_\oplus}$ such that $Pr(t_{R_\oplus}) = \sum_{t' \in R_\oplus \cap T(t)} Pr(t')$, and construct a random trial $X_{t_{R_\oplus}}$ whose success probability $Pr(X_{t_{R_\oplus}} = 1) = Pr(t_{R_\oplus})$.

Let $X_1,\ldots,X_n$ be the resulting trials. Since the independent tuples and rule-tuples in $T(t)$ are independent and their membership probabilities vary in general, $X_1,\ldots,X_n$ are independent and have unequal success probability values. They are Poisson trials. Let $X = \sum_{i=1}^{n} X_i$. Then, $Pr(T(t), j) = Pr(X = j)$ $(0 \leq j \leq n)$ where $Pr(X = j)$ is the probability of j successes. Thus, the probability that t is ranked the k-th is $Pr(t,k) = Pr(t)Pr(X = k-1)$. Moreover, the top-$k$ probability of t is given by $Pr^k(t) = Pr(t)Pr(X < k)$.

X follows the Poisson binomial distribution. Therefore, $Pr(t,k)$ also follows the Poisson binomial distribution, and $Pr^k(t)$ follows the cumulative distribution function of $Pr(t,k)$.

In a Poisson binomial distribution X, the probability density of X is unimodal (i.e., first increasing then decreasing), and attains its maximum at $\mu = E[X]$ [183]. Therefore, when the query parameter k varies from 1 to $|T(t)|+1$, $Pr(t,k)$ follows the similar trend.

Corollary 5.3 (Distribution of rank-k probability). *For a tuple $t \in T$,*

1. $Pr(t,k) \begin{cases} = 0, & \text{if } k > |T(t)|+1; \\ < Pr(t,k+1), & \text{if } k \leq \mu - 1; \\ > Pr(t,k+1), & \text{if } k \geq \mu. \end{cases}$
2. $\arg\max_{j=1}^{|T(t)|+1} Pr(t,j) = \mu + 1$, *where* $\mu = \sum_{t' \in T(t)} Pr(t')$. ∎

5.4.2 A General Stopping Condition

Corollary 5.3 shows that, given a tuple t and its compressed dominant set $T(t)$, the most possible ranks of t are around $\mu + 1$. In other words, if $k \ll \mu + 1$, then the top-k probability of t is small. Now, let us use this property to derive a general stopping condition for query answering algorithms progressively reading tuples in the ranking order. That is, once the stopping condition holds, all unread tuples that cannot satisfy the query can be pruned. The stopping condition is independent from the number of tuples in the data set, and dependent on only the query parameter k and the probability threshold p.

Theorem 5.9 (A General Stopping Condition). *Given a top-k query $Q^k(f)$ and probability threshold p, for a tuple $t \in T$, let $\mu = \sum_{t' \in T(t)} Pr(t')$. Then, $Pr^k(t) < p$ if $\mu \geq k + \ln\frac{1}{p} + \sqrt{\ln^2\frac{1}{p} + 2k\ln\frac{1}{p}}$.*

Proof. To prove Theorem 5.9, we need Theorem 4.2 in [184].

Lemma 5.1 (Chernoff Bound of Poisson Trials [184]). *Let $X_1, \ldots, X_n$ be independent Poisson trials such that, for $1 \leq i \leq n$, $Pr[X_i = 1] = p_i$, where $0 < p_i < 1$. Then, for $X = \sum_{i=1}^{n} X_i$, $\mu = E[X] = \sum_{i=1}^{n} p_i$, and $0 < \varepsilon \leq 1$, we have*

$$Pr[X < (1-\varepsilon)\mu] < e^{-\frac{\mu\varepsilon^2}{2}}.$$

As discussed in Section 5.4.1, we can construct a set of Poisson trials corresponding to the tuples in $T(t)$ such that, for each tuple or rule-tuple $t' \in T(t)$, there is a corresponding trial whose success probability is the same as $Pr(t')$. Moreover,

$$\sum_{j=0}^{k-1} Pr(T(t), j) = Pr[X < k]$$

For $0 < \varepsilon \leq 1$, inequality $Pr[X < k] \leq Pr[X < (1-\varepsilon)\mu]$ holds when

$$k \leq (1-\varepsilon)\mu \tag{5.1}$$

Using Lemma 5.1, we have

$$Pr[X < k] \le Pr[X < (1-\varepsilon)\mu] < e^{-\frac{\mu\varepsilon^2}{2}}$$

$Pr[X < k] < p$ holds if

$$e^{-\frac{\mu\varepsilon^2}{2}} \le p \tag{5.2}$$

Combining Inequality 5.1 and 5.2, we get $2\ln\frac{1}{p} \le \mu(1-\frac{k}{\mu})^2$. The inequality in Theorem 5.9 is the solution to the above inequality. ■

Since $\mu = \sum_{t' \in T(t)} Pr(t')$, the μ value is monotonically increasing if tuples are sorted in the ranking order. Using Theorem 5.9 an algorithm can stop and avoid retrieving further tuples in the rear of the sorted list if the μ value of the current tuple satisfies the condition in Theorem 5.9.

The value of parameter k is typically set to much smaller than the number of tuples in the whole data set. Moreover, since a user is interested in the tuples with a high probability to be ranked in top-k, the probability threshold p is often not too small. Consequently, μ is often a small value. For example, if $k = 100$, $p = 0.3$, then the stopping condition is $\mu \ge 117$.

In the experiments, we show in Figure 5.4 that the exact algorithm and the sampling algorithm stop close to the general stopping condition. The results verify the tightness of the stopping condition.

5.4.3 A Poisson Approximation Based Method

When the success probability is small and the number of Poisson trials is large, Poisson binomial distribution can be approximated well by Poisson distribution [185].

For a set of Poisson trials $X_1, \ldots, X_n$ such that $Pr(X_i = 1) = p_i$, let $X = \sum_{i=1}^{n} X_i$. X follows a Poisson binomial distribution. Let $\mu = E[X] = \sum_{i=0}^{n} p_i$. The probability of $X = k$ can be approximated by $Pr(X = k) \approx f(k;\mu) = \frac{\mu^k}{k!}e^{-\mu}$, where $f(k;\mu)$ is the Poisson probability mass function. Thus, the probability of $X < k$ can be approximated by $Pr(X < k) \approx F(k;\mu) = \frac{\Gamma(\lfloor k+1 \rfloor, \mu)}{\lfloor k \rfloor !}$, where $F(k;\mu)$ is the cumulative distribution function corresponding to $f(k;\mu)$, and $\Gamma(x,y) = \int_y^\infty t^{x-1}e^{-t}dt$ is the upper incomplete gamma function. Theoretically, Le Cam [186] showed that the quality of the approximation has the upper bound

$$\sup_{0 \le l \le n} \left| \sum_{k=0}^{l} Pr(X = k) - \sum_{k=0}^{l} f(k;\mu) \right| \le 9 \max_i \{p_i\}.$$

The above upper bound depends on only the maximum success probability in the Poisson trials. In the worst case where $\max_i\{p_i\} = 1$, the error bound is very loose. However, our experimental results (Figure 5.6) show that the Poisson approximation method achieves very good approximation quality in practice.

To use Poisson approximation to evaluate a top-k query $Q^k(f)$, we scan the tuples in T in the ranking order. The sum of membership probabilities of the scanned tuples is maintained in μ. Moreover, for each generation rule R, let R_{left} be the set of tuples in R that are already scanned. Correspondingly, let μ_R be the sum of membership probabilities of the tuples in R_{left}.

When a tuple t is scanned, if t is an independent tuple, then the top-k probability of t can be estimated using $Pr(t)F(k-1;\mu) = Pr(t)\frac{\Gamma(k,\mu)}{(k-1)!}$. If t belongs to a generation rule R, then the top-k probability of t can be estimated by $Pr(t)F(k-1;\mu') = Pr(t)\frac{\Gamma(k,\mu')}{(k-1)!}$, where $\mu' = \mu - \mu_R$. t is output if the estimated probability $Pr^k(t)$ passes the probability threshold p. The scan stops when the general stopping condition in Theorem 5.9 is satisfied.

In the Poisson approximation based method, we need to maintain the running μ and μ_R for each open rule R. Thus, the space requirement of the Poisson approximation based method is $O(|\mathscr{R}|+1)$, where $\mathscr{R}$ is the set of generation rules. The time complexity is $O(n')$, where n' is the number of tuples read before the general stopping condition is satisfied, which depends on parameter k, probability threshold p and the probability distribution of the tuples and is independent from the size of the uncertain table.

5.5 Online Query Answering

Since probabilistic ranking queries involve several parameters, a user may be interested in how query results change as parameters vary. To support the interactive analysis, online query answering is highly desirable.

In this section, we develop *PRist+* (for probabilistic ranking lists), an index for online answering probabilistic ranking queries on uncertain data, which is compact in space and efficient in construction.

We first describe the index data structure *PRist* and present a simple construction algorithm. Then, we develop *PRist+*, an advanced version of *PRist* that can be constructed more efficiently and achieve almost as good performance as *PRist* in query answering.

5.5.1 The PRist *Index*

To answer probabilistic ranking queries, for a tuple $t \in T$, a rank parameter k and a probability threshold p, we often need to conduct the following two types of checking operations.

- *Top-k probability checking*: is the top-k probability of t at least p?
- *p-rank checking*: is the p-rank of t at most k?

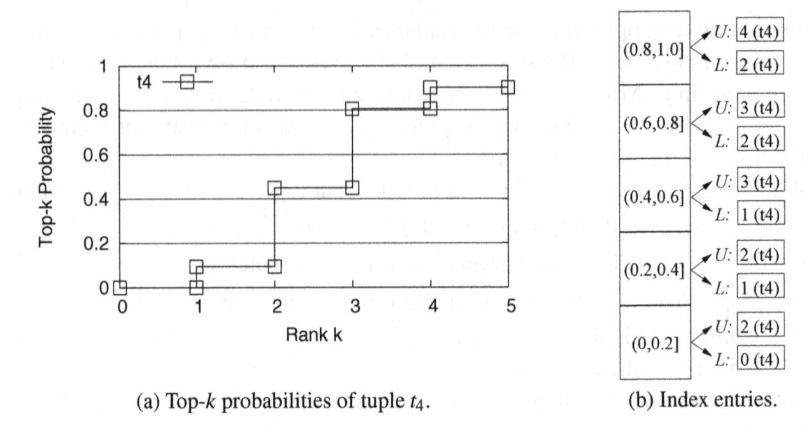

(a) Top-k probabilities of tuple t_4. (b) Index entries.

Fig. 5.2 The index entries in *PRist* for tuple t_4.

To support online query answering, we need to index the top-k probabilities and the p-ranks of tuples so that the checking operations can be conducted efficiently. Instead of building two different indexes for top-k probabilities and p-ranks separately, can we kill two birds with one stone?

One critical observation is that, for a tuple, the top-k probabilities and the p-ranks can be derived from each other. We propose *PRist*, a list of probability intervals, to store the rank information for tuples.

Example 5.4 (Indexing top-k probabilities). Let us consider how to index the uncertain tuples in Table 2.2.

Figure 5.2(a) shows the top-k probabilities of tuple t_4 with respect to different values of k. Interestingly, it can also be used to retrieve the p-rank of t_4: for a given probability p, we can draw a horizontal line for top-k probability $= p$, and then check where the horizontal line cut the curve in Figure 5.2(a). The point right below the horizontal line gives the answer k.

Storing the top-k probabilities for all possible k can be very costly, since k is in range $[1,m]$ where m is the number of rules in the data set. To save space, we can divide the domain of top-k probabilities $(0,1]$ into h prob-intervals.

In Figure 5.2(a), we partition the probability range $(0,1]$ to 5 prob-intervals: $(0,0.2]$, $(0.2,0.4]$, $(0.4,0.6]$, $(0.6,0.8]$, and $(0.8,1.0]$. For each interval, we record for each tuple a lower bound and an upper bound of the ranks whose corresponding top-k probabilities lie in the prob-interval.

For example, the top-1 and top-2 probabilities of t_4 are 0.0945 and 0.45, respectively. Therefore, a lower bound and an upper bound of ranks of t_4 in interval $(0,0.2]$ are 0 and 2, respectively. Specifically, we define the top-0 probability of any tuples as 0. Moreover, we choose rank $k = 2$ as the upper bound of the 0.2-rank of t_4, since the top-2 probability of t_4 is greater than 0.2 and all top-i probabilities of t_4 for $i < 2$ are smaller than 0.2. The lower bound of the 0-rank probability of t_4 is chosen similarly.

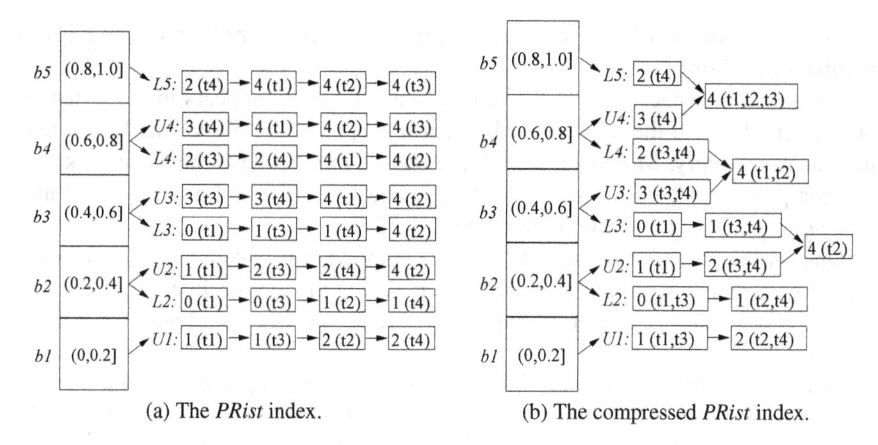

Fig. 5.3 A *PRist* index for the uncertain tuples in Table 2.2.

To store the bounds in interval $(0,0.2]$, *we maintain two lists associated with the interval. The L-list stores a lower bound for each tuple, and the U-list stores an upper bound for each tuple. Therefore, for* t_4, *lower bound* 0 *is inserted into the L-list of* $(0,0.2]$, *and upper bound* 2 *is inserted into the U-list of* $(0,0.2]$.

The lower and upper bounds of ranks for other prob-intervals can be computed in the same way. Finally, the entry of t_4 *in a* PRist *is shown in Figure 5.2(b).* ■

Formally, given a set of uncertain tuples T and a *granularity parameter* $h > 0$, a *PRist* index for T contains a set of *prob-intervals* $\{b_1, \ldots, b_h\}$, where $b_i = (\frac{i-1}{h}, \frac{i}{h}]$ $(1 \leq i \leq h)$.

Each prob-interval b_i $(2 \leq i \leq h-1)$ is associated with two lists: a U-list and an L-list.

An entry in the U-list of b_i corresponds to a tuple t and consists of two items: tuple id t and an *upper rank* of t in b_i, denoted by $t.U_i$, such that one of the following holds: (1) $Pr^{t.U_i}(t) > \frac{i}{h}$ and $Pr^{t.U_i - 1} \leq \frac{i}{h}$ when $Pr^m(t) > \frac{i}{h}$; or (2) $t.U_i = m$ when $Pr^m(t) \leq \frac{i}{h}$. Each tuple $t \in T$ has an entry in the U-list. All entries in the U-list are sorted in ascending order of the upper ranks.

An entry in the L-list of b_i corresponds to a tuple t and consists of two items: tuple id t and a *lower rank* of t in b_i, denoted by $t.L_i$, such that one of the following holds: (1) $Pr^{t.L_i}(t) \leq \frac{i-1}{h}$ and $Pr^{t.L_i+1}(t) > \frac{i-1}{h}$ when $Pr^m(t) > \frac{i-1}{h}$; or (2) $t.L_i = m$ when $Pr^m(t) \leq \frac{i-1}{h}$. Each tuple $t \in T$ has an entry in the L-list. All entries in the L-list are sorted in ascending order of the lower ranks.

Prob-interval b_1 is associated with only a U-list but no L-list. Prob-interval b_h is associated with only an L-list but no U-list. For any tuple $t \in T$, $t.L_1 = 0$ and $t.U_h = m$. The reason is that the lower ranks for entries in b_1 are always 0 and the upper ranks for entries in b_h are always m. Those two lists can be omitted.

*Example 5.5 (*PRist*). For tuples* $t_1, \cdots, t_4$ *in Table 2.2, whose top-k probabilities are plotted in Figure 2.5, the* PRist *index is shown in Figure 5.3(a).* ■

The space cost of a *PRist* is $O(2hn)$, where n is the number of tuples and h is the number of prob-intervals.

To reduce the space cost of *PRist*, if an entry $(o, rank)$ appears in the U-list of prob-interval b_i and the L-list of prob-interval b_{i+1} $(1 \leq i < h)$, we can let the two lists share the entry. Moreover, if multiple entries in a list have the same rank, we can compress those entries into one which carries one rank and multiple tuple ids. A *compressed PRist* index is shown in Figure 5.3(b).

One reason that compressed *PRist* can achieve non-trivial saving is that, for a tuple t with membership probability $Pr(t)$, if $Pr(t)$ falls in prob-interval b_i, the upper and lower ranks of t in prob-intervals $b_{i+1}, \ldots, b_h$ are identical and thus can be shared.

In addition, *PRist* can be tailored according to different application requirements. For example, in some scenarios, users may be only interested in the probabilistic ranking queries with a small rank parameter k and a large probability threshold p. In such a case, given a maximum rank parameter value k_{max} and a minimum probability threshold value p_{min} in users' queries, only the prob-intervals $(\frac{i-1}{h}, \frac{i}{h}]$ with $\frac{i}{h} > p_{min}$ need to be stored. For each prob-interval, the U-list and L-list only contain the tuples whose upper ranks are at most k_{max}.

We can build a *PRist* index in two steps. In the first step, we compute the top-k probabilities of all tuples using the methods in Section 5.2. The time complexity of computing the top-i probabilities of all tuples for all $1 \leq i \leq m$ is $O(m^2 n)$, where m is the number of rules in T and n is the number of tuples in T.

In the second step, we construct a set of prob-intervals and compute the U-lists and L-lists. For each tuple t, we scan its top-i $(1 \leq i \leq m)$ probabilities and fill up the upper rank $t.U_j$ and lower rank $t.L_j$ for each prob-interval b_j $(1 \leq j \leq h)$. Since there are n tuples and m ranks, the time cost is $O(mn)$. We sort the entries in the U-lists and the L-lists. Since there are $2h$ lists, each of n entries, the total time cost of sorting is $O(2hn \log n)$.

The overall time complexity of the basic construction algorithm is $O(m^2 n + mn + 2hn \log n) = O(m^2 n + 2hn \log n)$.

5.5.2 *Query Evaluation based on* PRist

The query evaluation based on the *PRist* index follows three steps.

1. **Bounding**: for each tuple in question, we derive an upper bound and a lower bound of the measure of interest;
2. **Pruning and validating**: The upper and lower bound derived in the bounding phase are used to prune or validate tuples.
3. **Evaluating**: for those tuples that cannot be pruned or validated, the exact top-k probabilities are calculated. Then, the tuples are evaluated with respect to the query.

5.5.2.1 Answering PT-k Queries

The following example illustrates how to evaluate PT-k queries.

Example 5.6 (Answering PT-k queries). Consider the uncertain tuples indexed in Figure 5.3 again, and a PT-k query with $k = 3$ and $p = 0.45$.

To find the tuples satisfying the query, we only need to look at the prob-interval containing $p = 0.45$, which is $b_3 = (0.4, 0.6]$. In the U-list of b_3, we find that $t_3.U_3 = 3$ and $t_4.U_3 = 3$, which means that $Pr^3(t_3) > 0.6$ and $Pr^3(t_4) > 0.6$. Therefore, t_3 and t_4 can be added into the answer set without calculating their exact top-k probabilities. In the L-list of b_3, we find that $t_2.L_3 = 4$, which means $Pr^4(t_2) \leq 0.4$. Therefore, t_2 can be pruned.

Thus, only the top-3 probability of t_1 needs to be calculated in order to further verify if t_1 is an answer to the query. Since $Pr^3(t_1) = 0.5$, it can be added into the answer set. The final answer is $\{t_1, t_3, t_4\}$. ■

The three steps for PT-k query evaluation are as follows.

Step 1: Bounding We use Corollary 5.4 to determine whether the top-k probability of t lies in b_i.

Corollary 5.4 (Bounding top-k probabilities). *Let T be a set of uncertain tuples indexed by* PRist *with granularity parameter h. For a tuple $t \in T$ and a positive integer k, if b_i $(1 \leq i \leq h)$ is the prob-interval such that $t.L_i < k < t.U_i$, then $\frac{i-1}{h} < Pr^k(t) \leq \frac{i}{h}$.*

Proof. According to the definition of *PRist*, we have $Pr^{t.L_i}(t) \leq \frac{i-1}{h}$ and $Pr^{t.L_i+1}(t) > \frac{i-1}{h}$. Since $k > t.L_i$, $Pr^k(t) \geq Pr^{t.L_i+1}(t) > \frac{i-1}{h}$. On the other hand, $Pr^{t.U_i}(t) > \frac{i}{h}$ and $Pr^{t.U_i-1}(t) \leq \frac{i}{h}$. Since $k < t.U_i$, we have $Pr^k(t) \leq Pr^{t.U_i-1}(t) \leq \frac{i}{h}$. ■

Step2: Pruning and Validating A tuple may be pruned or validated by checking its lower rank L_i or upper rank U_i in the prob-interval containing the probability threshold, as stated in Theorem 5.10.

Theorem 5.10 (Answering PT-k queries). *Let T be a set of uncertain tuples indexed by* PRist *with granularity parameter h. For a tuple $t \in T$ and a PT-k query Q^k_f with probability threshold $p \in b_i = (\frac{i-1}{h}, \frac{i}{h}]$:*

1. Pruning: *if $t.L_i > k$, then $Pr^k(t) < p$;*
2. Validating: *if $t.U_i \leq k$, then $Pr^k(t) > p$.*

Proof. According to the definition of *PRist*, we have $Pr^{t.L_i}(t) \leq \frac{i-1}{h}$ and $Pr^{t.L_i+1}(t) > \frac{i-1}{h}$. Therefore, if $t.L_i > k$, $Pr^k(t) \leq Pr^{t.L_i}(t) \leq \frac{i-1}{h} < p$. Moreover, according to the definition of *PRist*, $Pr^{t.U_i}(t) > \frac{i}{h}$ and $Pr^{t.U_i-1}(t) \leq \frac{i}{h}$. Thus, if $t.U_i \leq k$, then $Pr^k(t) \geq Pr^{t.U_i}(t) > \frac{i}{h} > p$. ■

Step 3: Evaluating We only need to evaluate the exact top-k probabilities for those tuples whose top-k probabilities falling into the prob-interval containing the probability threshold p, and which cannot be validated or pruned by Theorem 5.10.

In the U-list of the prob-interval containing the probability threshold, finding the tuples whose upper ranks are less than or equal to k requires $O(\log n)$ time, where n is the number of tuples in T. Similarly, in the L-list, finding the tuples whose lower ranks are larger than k also takes $O(\log n)$ time. Let d be the number of tuples that cannot be pruned or validated. Computing the top-k probabilities of those tuples requires $O(kmd)$ time, where m is the number of rules in T.

5.5.2.2 Answering Top-(k,l) Queries

Example 5.7 (Answering Top-(k,l) queries). Consider the uncertain tuples in Table 2.2 which are indexed in the PRist *in Figure 5.3. To answer a top-(k,l) with $k=3$ and $l=2$, we scan each prob-interval in Figure 5.3(a) from top down.*

In prob-interval $b_5 = (0.8, 1.0]$, we find $t_4.L_5 = 2$, which means $Pr^2(t_4) \leq 0.8$ and $Pr^3(t_4) > 0.8$. At the same time, $t_j.L_5 = 4$ for $1 \leq j \leq 3$, which means $Pr^4(t_j) \leq 0.8$ for $1 \leq j \leq 3$. Since for any tuple t, we have $Pr^3(t) \leq Pr^4(t)$, it is clear that $Pr^3(t_4) < Pr^3(t_j)$ $(1 \leq j \leq 3)$. Thus, t_4 is one of the answers to the query.

Using the similar procedure, we scan prob-interval b_4 and find that t_3 is another answer. Since the query asks for the top-2 results, the query answering procedure stops. ■

To answer a top-(k,l) query Q, we want to scan the tuples in the descending order of their top-k probabilities. However, *PRist* does not store any exact top-k probabilities. We scan the prob-intervals in the top-down manner instead. For each prob-interval, we retrieve the tuples whose top-k probabilities lie in the prob-interval. Obviously, for two prob-intervals b_i and b_j $(i > j)$, the top-k probabilities falling in b_i is always greater than the top-k probabilities in b_j.

The query answering algorithm proceeds in three steps.

Step 1: Bounding For a prob-interval b_i and a tuple t, we use Corollary 5.4 to determine if the top-k probability of t lies in b_i.

Step 2: Pruning and Validating We scan the prob-intervals from top down, and use a buffer B to store the tuples whose top-k probabilities lie in the scanned prob-intervals. The tuples not in B have smaller top-k probabilities than those in B. A tuple may be pruned or validated according to whether it is in B and the number of tuples in B.

Theorem 5.11 (Answering top-(k,l) queries). *Let T be a set of uncertain tuples indexed by* PRist *with granularity parameter h. Let $b_h, b_{h-1} \ldots, b_i$ $(1 \leq i \leq h)$ be the prob-intervals that have been scanned and B contain all tuples whose top-k probabilities lying in $b_h, b_{h-1} \ldots, b_i$, then*

1. Pruning*: if $|B| \geq l$ and $t \notin B$, then t is not an answer to the top-(k,l) query;*
2. Validating*: if $|B| \leq l$ and $t \in B$, then t is an answer to the top-(k,l) query.*

Proof. We only need to prove that for any tuple $t_x \in B$ and $t_y \notin B$, $Pr^k(t_x) < Pr^k(t_y)$. Since $t_x \in B$, we have $Pr^k(t_x) > \frac{i-1}{h}$. Moreover, $Pr^k(t_x) \leq \frac{i-1}{h}$ because $t_y \notin B$. Therefore, we have $Pr^k(t_x) < Pr^k(t_y)$. ■

Step 3: Evaluating If, before prob-interval b_i is scanned, B has l' tuples such that $l' < l$, but after b_i is scanned, B has l or more tuples, we enter the evaluating step. If B has exactly l tuples, we do not need to do anything. However, if B has more than l tuples, we need to evaluate the exact top-k probabilities of those tuples added to B from b_i, and include into B only the top-$(l - l')$ tuples of the largest top-k probabilities.

When a tuple can be added into the answer set without the evaluating step, the time to retrieve the tuple is constant. There are at most l such tuples. Only the top-k probabilities of the tuples in the last prob-interval need to be evaluated. Let d be the number of such tuples. Then, the time complexity of evaluating those tuples is $O(kmd)$, as computing the top-k probabilities of those tuples requires $O(kmd)$ time. The overall time complexity of query answering is $O(l + kmd)$.

5.5.2.3 Answering Top-(p,l) Queries

Example 5.8 (Answering top-(p,l) queries). Consider the uncertain tuples in Figure 5.3 again. How can we answer the top-(p,l) query with $p = 0.5$ and $l = 2$?

We only need to check prob-interval $b_3 = (0.4, 0.6]$ that contains p. First, the top-2 highest ranks in $b_3.U_3$ are 3 for tuples t_3 and t_4, which means $MR_{0.5}(t_3) \leq 3$ and $MR_{0.5}(t_4) \leq 3$. Therefore, we can set 3 as the upper bound of the 0.5-ranks, and any tuple t with $MR_{0.5} > 3$ cannot be the answer to the query.

Second, we scan list $b_3.L_3$ from the beginning. t_1 is added into a buffer B, since $t_1.L_3 < 3$ and $MR_{0.5}$ might be smaller than or equal to 3. t_3 and t_4 are added into B due to the same reason. Then, when scanning t_2, we find $t_2.L_3 = 4$, which means $MR_{0.5}(t_2) \geq MR_{0.4}(t_2) \geq 4$. Thus, t_2 and any tuples following t_2 in the list (in this example, no other tuples) can be pruned.

Last, the 0.5-ranks of the tuples in B (i.e., t_1, t_3, and t_4) are calculated, and the tuples with the top-2 highest 0.5-ranks are returned as the answer to the query, which are t_1 and t_3. ■

To answer a top-(p,l) query, the three steps work as follows.

Step 1: Bounding The p-rank of a tuple can be bounded using the following rule.

Theorem 5.12 (Bounding p-ranks). *Let T be a set of uncertain tuples indexed by* PRist *with granularity parameter h. For a tuple $t \in S$ and $p \in (0,1]$, let b_i $(1 \leq i \leq h)$ be the prob-interval such that $p \in b_i$, i.e., $\frac{i-1}{h} < p \leq \frac{i}{h}$. Then, $t.L_i \leq MR_p(t) \leq t.U_i$.*

Proof. To prove Theorem 5.12, we need the following two lemmas.

Lemma 5.2 (Monotonicity). *Let t be an uncertain tuple, and p_1 and p_2 be two real values in $(0,1]$. If $p_1 \leq p_2$, then $MR_{p_1}(t) \leq MR_{p_2}(t)$.*

Proof. Let $MR_{p_1}(t) = k_1$. Then, $Pr^{k_1}(t) \geq p_1$. Moreover, for any $k < k_1$, $Pr^k(t) < p_1$. Since $p_1 \leq p_2$, for any $k < k_1$, $Pr^k(t) < p_2$. Therefore, $MR_{p_2}(t) \geq k_1 = MR_{p_1}(t)$.

Lemma 5.3. *Let t be an uncertain tuple, k be a positive integer, and p be a real values in $(0,1]$. If $Pr^k(t) \leq p$, then $MR_p(t) \geq k$.*

Proof. Since $Pr^k(t) \leq p$, for any $x \leq k$, $Pr^x(t) \leq p$. Thus, $MR_p(t) \geq k$.

To proof Theorem 5.12, We first prove $t.L_i < MR_p(t)$. Following the definition of *PRist*, we have $Pr^{t.L_i}(t) \leq \frac{i-1}{h}$. Thus, $MR_{\frac{i-1}{h}}(t) \geq t.L_i$, which follows from Lemma 5.3. Moreover, since $\frac{i-1}{h} < p$, we have $MR_{\frac{i-1}{h}}(t) \leq MR_p(t)$, which follows from Lemma 5.2. Thus, $t.L_i < MR_p(t)$.

Similarly, $MR_p(t) \leq t.U_i$ can be proved. ■

Step 2: Pruning and Validating Given a top-(p,l) query, let b_i be the prob-interval containing p. We use the l-th rank k in list $b_i.U$ as a pruning condition. Any tuple t' whose lower rank in b_i is at least k can be pruned. Moreover, for any tuple t, if the upper rank of t in b_i is not greater than l lower ranks of other tuples in b_i, then t can be validated.

Theorem 5.13 (Answering Top-(p,l) queries). *Let T be a set of uncertain tuples indexed by* PRist *with granularity parameter h. For a tuple $t \in S$, a top-(p,l) query, and prob-interval b_i such that $p \in b_i$:*

1. Pruning*: Let $T_1 = \{t_x | t_x.U_i \leq t.U_i \wedge t_x \in T\}$. If $|T_1| \geq l$, then any tuple t' such that $t'.L_i \geq t.U_i$ is not an answer to the query;*
2. Validating*: Let $T_2 = \{t_y | t_y.L_i \leq t.U_i \wedge t_y \in T\}$. If $|T_2| \leq l$, then t is an answer to the query.*

Proof. To prove the first item, we only need to show that there are at least l tuples whose p-ranks are smaller than $MR_p(t')$. For any tuple $t_x \in T_1$, we have $MR_p(t_x) \leq t_x.U_i \leq t.U_i$, which is guaranteed by Theorem 5.12. Similarly, for any tuple t' such that $t'.L_i \geq t.U_i$, we have $MR_p(t') \geq t.U_i$. Therefore, $MR_p(t_x) \leq MR_p(t')$. Since $|T_1| \geq l$, there are at least l tuples like t_x. So the first item holds.

To prove the second item, we only need to show that there are fewer than l tuples whose p-ranks are smaller than $MR_p(t)$. For any tuple $t' \notin T_2$, since $t'.L_i > t.U_i$, $MR_p(t') > MR_p(t)$. Therefore, only the tuples in T_2 may have smaller p-ranks than $MR_p(t)$. Since $|T_2| \leq l$, the second item holds. ■

Step 3: Evaluating For any tuple that can neither be validated nor be pruned, we need to calculate their exact p-ranks. To calculate the p-rank of tuple t, we calculate the top-1, top-2, $\cdots$, top-i probabilities for each rank i until we find the first rank k such that $Pr^k(t) \geq p$.

The complexity is analyzed as follows. Let d_1 be number of tuples that cannot be pruned in the pruning and validation phase. In the evaluating phase, the p-ranks of the tuples that cannot be pruned or validated are calculated. For each such tuple, its p-rank can be m in the worst case, where m is the number of rules in T. Then, calculating the p-rank for each tuple takes $O(m^2)$ time. Let d_2 be the number of tuples that cannot be pruned or validated. The evaluating step takes $O(m^2 d_2)$ time. The overall complexity is $O(d_1 + m^2 d_2)$.

5.5.3 PRist+ and a Fast Construction Algorithm

We can reduce the construction time of *PRist* by bounding top-k probabilities using the binomial distribution.

Consider a tuple $t \in T$ and its compressed dominant set $T(t)$. If there is any tuple or generation rule-tuple with probability 1, we can remove the tuple from $T(t)$, and compute the top-$(k-1)$ probability of t. Thus, we can assume that the membership probability of any tuple or rule-tuple in $T(t)$ is smaller than 1.

Theorem 5.14 (Bounding the probability). *For a tuple $t \in T$, let $T(t)$ be the compressed dominant set of t. Then,*

$$F(k;N,p_{max}) \leq \sum_{0 \leq j \leq k} Pr(T(t),j) \leq F(k;N,p_{min}) \tag{5.3}$$

where p_{max} and p_{min} are the greatest and the smallest probabilities of the tuples/rule-tuples in $T(t)$ $(0 < p_{min} \leq p_{max} < 1)$, N is the number of tuples/rule-tuples in $T(t)$, and F is the cumulative distribution function of the binomial distribution.

Proof. We first prove the left side of Inequality 5.3. For a tuple set S, let $Pr(S, \leq k)$ denote $\sum_{0 \leq j \leq k} Pr(S,j)$. For any tuple $t' \in T(t)$, $Pr(t') \leq p_{max}$.

Consider tuple set $S = T(t) - \{t'\}$ and $T'(t) = S + t_{max}$ where $Pr(t_{max}) = p_{max}$. From Theorem 5.2,

$$Pr(T(t), \leq k) = Pr(t')Pr(S, \leq k-1) + (1 - Pr(t'))Pr(S, \leq k);$$

and

$$Pr(T'(t), \leq k) = Pr(t_{max})Pr(S, \leq k-1) + (1 - Pr(t_{max}))Pr(S, \leq k).$$

Then,

$$Pr(T(t), \leq k) - Pr(T'(t), \leq k) = [Pr(t') - Pr(t_{max})] \times [Pr(S, \leq k-1) - Pr(S, \leq k)].$$

Since $Pr(t') \leq Pr(t_{max})$ and $Pr(S, \leq k-1) \leq Pr(S, \leq k)$, we have $Pr(T(t), \leq k) \geq Pr(T'(t), \leq k)$.

By replacing each tuple/rule-tuple in $T(t)$ with t_{max}, we obtain a set of tuples with the same probability p_{max}, whose subset probabilities follows the binomial distribution $F(k;N,p_{max})$. Thus, the left side of Inequality 5.3 is proved.

The right side of Inequality 5.3 can be proved similarly. ■

Moreover, Hoeffding [183] gave the following bound.

Theorem 5.15 (Extrema [183]). *For a tuple $t \in T$ and its compressed dominant set $T(t)$, let $\mu = \sum_{t' \in T(t)} Pr(t' \prec_f t)$. Then,*

1. *$\sum_{j=0}^{k} Pr(T(t),j) \leq F(k;N,\frac{\mu}{N})$ when $0 \leq k \leq \mu - 1$; and*
2. *$\sum_{j=0}^{k} Pr(T(t),j) \geq F(k;N,\frac{\mu}{N})$ when $\mu \leq k \leq N$,*

where N is the number of tuples and rule-tuples in $T(t)$, and F is the cumulative distribution function of the binomial distribution. ■

Based on Theorems 5.14 and 5.15, we derive the following bound for the top-k probability of tuple $t \in T$.

Theorem 5.16 (Bounds of top-k probabilities). *For a tuple $t \in T$, the top-k probability of t satisfies*

1. $Pr(t)F(k-1;N,p_{max}) \leq Pr^k(t) \leq Pr(t)F(k-1;N,\frac{\mu}{N})$ *for* $1 \leq k \leq \mu$;
2. $Pr(t)F(k-1;N,\frac{\mu}{N}) \leq Pr^k(t) \leq Pr(t)F(k-1;N,p_{min})$ *for* $\mu+1 \leq k \leq N+1$.

Proof. The theorem holds following from Theorems 5.14 and 5.15. ■

Example 5.9 (Bounding the probability). Consider the uncertain tuples in Table 2.2 again. $T(t_3) = \{t_1,t_2\}$. *The expected number of tuples in* $T(t_3)$ *is* $\mu = Pr(t_1) + Pr(t_2) = 0.8$. *The number of rules in* $T(t_3)$ *is 2.*

Consider the top-2 probability of t_3. *Since* $2 > \mu$, $Pr^2(t_3) \geq Pr(t_3)F(1;2,0.4) = 0.7 \times 0.84 = 0.588$. *Since* $\min\{Pr(t_1),Pr(t_2)\} = 0.3$, $Pr^2(t_3) \leq Pr(t_3)F(1;2,0.3) = 0.7 \times 0.91 = 0.637$. *Therefore,* $Pr^2(t_3)$ *is bounded in range* $[0.588, 0.637]$. ■

Since the cumulative probability distribution of the binomial distribution is easier to calculate than top-k probabilities, we propose *PRist+*, a variant of *PRist* using the binomial distribution bounding technique.

The only difference between *PRist+* and *PRist* is the upper and lower ranks in the U-lists and L-lists. In *PRist+*, we compute the upper and lower bounds of the top-k probabilities of tuples using the binomial distributions. Then, the upper and lower ranks are derived from the upper and lower bounds of the top-k probabilities.

Take the U-list in prob-interval b_i as an example. In *PRist*, an entry in the U-list consists of the tuple id t and the upper rank $t.U_i$, such that $Pr^{t.U_i}(t) > \frac{i}{h}$ (if $Pr^m(t) > \frac{i}{h}$). Once the top-i probabilities of all ranks $1 \leq i \leq m$ for t are computed, $t.U_i$ can be obtained by one scan.

In *PRist+*, we store the upper rank $t.U_i$ as the smallest rank x such that the lower bound of $Pr^x(t)$ is greater than $\frac{i}{h}$. Following with Theorem 5.16, the upper rank can be calculated by $x = F^{-1}(\frac{i}{h \cdot Pr(t)}, N, \frac{\mu}{N}) + 1$ or $x = F^{-1}(\frac{i}{h \cdot Pr(t)}, N, p_{max}) + 1$, where F^{-1} is the binomial inverse cumulative distribution function. The lower rank of t can be obtained similarly using Theorem 5.16.

Computing the upper and lower ranks for t in b_i requires $O(1)$ time. Thus, the overall complexity of computing the upper and lower ranks of all tuples in all prob-intervals is $O(2hn)$, where n is the total number of tuples. The complexity of sorting the bound lists is $O(2hn\log n)$. The overall time complexity of constructing a *PRist+* index is $O(2hn + 2hn\log n) = O(hn\log n)$.

Clearly, the construction time of *PRist+* is much lower than *PRist*. The tradeoff is that the rank bounds in *PRist+* is looser than *PRist*. As will be shown in the next section, all query answering methods on *PRist* can be applied on *PRist+*. The looser rank bounds in *PRist+* does not affect the accuracy of the answers. They only make a very minor difference in query answering time in our experiments.

5.6 Experimental Results

We conduct a systematic empirical study using a real data set and some synthetic data sets on a PC with a 3.0 GHz Pentium 4 CPU, 1.0 GB main memory, and a 160 GB hard disk, running the Microsoft Windows XP Professional Edition operating system. Our algorithms were implemented in Microsoft Visual C++ V6.0.

Tuple	*R*1	*R*2	*R*3	*R*4	*R*5	*R*6	*R*7	*R*8	*R*9	*R*10	*R*11	*R*14	*R*18
Drifted days	435.8	341.7	335.7	323.9	284.7	266.8	259.5	240.4	233.6	233.3	232.6	230.9	229.3
Membership probability	0.8	0.8	0.8	0.6	0.8	0.8	0.4	0.15	0.8	0.7	0.8	0.6	0.8
Top-10 prob.	0.8	0.8	0.8	0.6	0.8	0.8	0.4	0.15	0.8	0.7	0.79	0.52	0.359

Table 5.2 Some tuples in the IIP Iceberg Sightings Database 2006.

Rank	1	2	3	4	5	6	7	8	9	10
Tuple	*R*1	*R*2	*R*3	*R*5	*R*6	*R*9	*R*9	*R*11	*R*11	*R*18
$Pr(t,j)$	0.8	0.64	0.512	0.348	0.328	0.258	0.224	0.234	0.158	0.163

Table 5.3 The answers to the *U-KRanks* query.

5.6.1 Results on IIP Iceberg Database

We use the International Ice Patrol (IIP) Iceberg Sightings Database[1] to examine the effectiveness of top-k queries on uncertain data in real applications. The International Ice Patrol (IIP) Iceberg Sightings Database collects information on iceberg activities in the North Atlantic. The mission is to monitor iceberg danger near the Grand Banks of Newfoundland by sighting icebergs (primarily through airborne Coast Guard reconnaissance missions and information from radar and satellites), plotting and predicting iceberg drift, and broadcasting all known ice to prevent icebergs threatening.

In the database, each sighting record contains the sighting date, sighting location (latitude and longitude), number of days drifted, etc. Among them, the number of days drifted is derived from the computational model of the IIP, which is crucial in

[1] `http://nsidc.org/data/g00807.html`

$k=5$			$k=20$		
RID	Top-5 prob.	# of Days Drifted	RID	Top-20 Prob.	# of Days Drifted
$R1$	0.8	435.8	$R1$	0.8	435.8
$R2$	0.8	341.7	$R2$	0.8	341.7
$R3$	0.8	335.7	$R3$	0.8	335.7
$R5$	0.8	284.7	$R5$	0.8	284.7
$R6$	0.61	266.8	$R6$	0.8	266.8
$R4$	0.6	323.9	$R9$	0.8	233.6
$R9$	0.22	233.6	$R11$	0.8	232.6
$R7$	0.17	259.5	$R18$	0.8	229.3
$R10$	0.09	233.3	$R23$	0.79	227.2
$R8$	0.05	240.4	$R33$	0.75	222.2

Table 5.4 Results of top-(k,l) queries on the IIP Iceberg Sighting Database (l=10).

determining the status of icebergs. It is interesting to find the icebergs drifting for a long period.

However, each sighting record in the database is associated with a confidence level according to the source of sighting, including: R/V (radar and visual), VIS (visual only), RAD(radar only), SAT-L(low earth orbit satellite), SAT-M (medium earth orbit satellite) and SAT-H (high earth orbit satellite). In order to quantify the confidence, we assign confidence values 0.8, 0.7, 0.6, 0.5, 0.4 and 0.3 to the above six confidence levels, respectively.

Moreover, generation rules are defined in the following way. For the sightings with the same time stamp, if the sighting locations are very close – differences in latitude and longitude are both smaller than 0.01 (i.e.,0.02 miles), they are considered referring to the same iceberg, and only one of the sightings is correct. All tuples involved in such a sighting form a multi-tuple rule. For a rule $R: t_{r_1} \oplus \cdots \oplus t_{r_m}$, $Pr(R)$ is set to the maximum confidence among the membership probability values of tuples in the rule. Then, the membership probability of a tuple is adjusted to $Pr(t_{r_l}) = \frac{conf(t_{r_l})}{\sum_{1 \leq i \leq m} conf(t_{r_i})} Pr(R)$ $(1 \leq l \leq m)$, where $conf(t_{r_l})$ is the confidence of t_{r_l}. After the above preprocessing, the database contains 4,231 tuples and 825 multi-tuple rules. The number of tuples involved in a rule varies from 2 to 10. We name the tuples in the number of drifted days descending order. For example, tuple $R1$ has the largest value and $R2$ has the second largest value on the attribute.

5.6.1.1 Comparing PT-k queries, U-Top$k$ queries and U-KRanks Queries

We apply a *PT-k query*, a *U-TopK query* and a *U-KRanks query* on the database by setting $k=10$ and $p=0.5$. The ranking order is the number of drifted days descending order. The PT-k query returns a set of 10 records $\{R1, R2, R3, R4, R5, R6, R9, R10, R11, R14\}$. The U-Top$k$ query returns a vector $\langle R1, R2, R3, R4, R5, R6, R7, R9, R10, R11\rangle$ with probability 0.0299. The U-KRanks query returns 10 tuples shown in Table 5.3. The probability values of the tuples at the corresponding ranks are also shown in the table. To understand the answers, in Table 5.2 we also list the member-

$p=0.5$			$p=0.7$		
RID	0.5-rank	# of Days	RID	0.7-rank	# of Days
$R1$	2	435.8	$R1$	2	435.8
$R2$	2	341.7	$R2$	2	341.7
$R3$	3	335.7	$R3$	3	335.7
$R4$	4	323.9	$R5$	5	284.7
$R5$	4	284.7	$R6$	6	266.8
$R6$	5	266.8	$R9$	7	233.6
$R9$	7	233.6	$R11$	9	232.6
$R10$	8	233.3	$R10$	10	233.3
$R11$	8	232.6	$R18$	13	229.3
$R14$	10	231.1	$R23$	15	227.2

Table 5.5 Results of Top-(p,l) queries on the IIP Iceberg Sighting Database (l=10).

ship probability values and the top-10 probability values of some tuples including the ones returned by the PT-k, U-Top$k$, and U-KRanks queries.

All tuples with a top-10 probability of at least 0.5 are returned by the PT-k query. The top-10 probability of $R14$ is higher than $R7$, but $R7$ is included in the answer of the U-Top$k$ query and $R14$ is missing. Moreover, the presence probability of the top-10 list returned by the U-Top$k$ query is quite low. Although it is the most probable top-10 tuple list, the low presence probability limits its usefulness and interestingness.

$R10$ and $R14$, whose top-10 probability values are high, are missing in the results of the U-KRanks query, since none of them is the most probable at any rank. Nevertheless, $R18$ is returned by the U-KRanks query at the 10-th position, though its top-10 probability is much lower than $R10$ and $R14$. Moreover, $R9$ and $R11$ each occupies two positions in the answer of the U-KRanks query.

The results clearly show that the *PT-k query* captures some important tuples missed by the *U-TopK query* and the *U-KRanks query*.

5.6.1.2 Answering Top-(k,l) Queries, PT-k Queries and Top-(p,l) Queries

Moreover, We conduct top-(k,l) queries, PT-k queries and top-(p,l) queries on the database.

Table 5.4 shows the results of a top-$(5,10)$ query and a top-$(20,10)$ query. To understand the answer, the top-5 probabilities, the top-20 probabilities, and the number of days drifted are also included in the table. Some records returned by the top-$(5,10)$ are not in the results of the top-$(20,10)$ query, such as $R4$, $R7$, $R10$ and $R8$. Moreover, $R11$, $R18$, $R23$ and $R33$ are returned by the top-$(20,10)$ query but are not in the results of the top-$(5,10)$ query. When k becomes larger, the records ranked relatively lower but with larger membership probabilities may become the answers. It is interesting to vary value k and compare the difference among the query results.

The results to a top-$(0.5,10)$ query and a top-$(0.7,10)$ query are listed in Table 5.5. The 0.5-ranks, 0.7-ranks and the number of days drifted are also included.

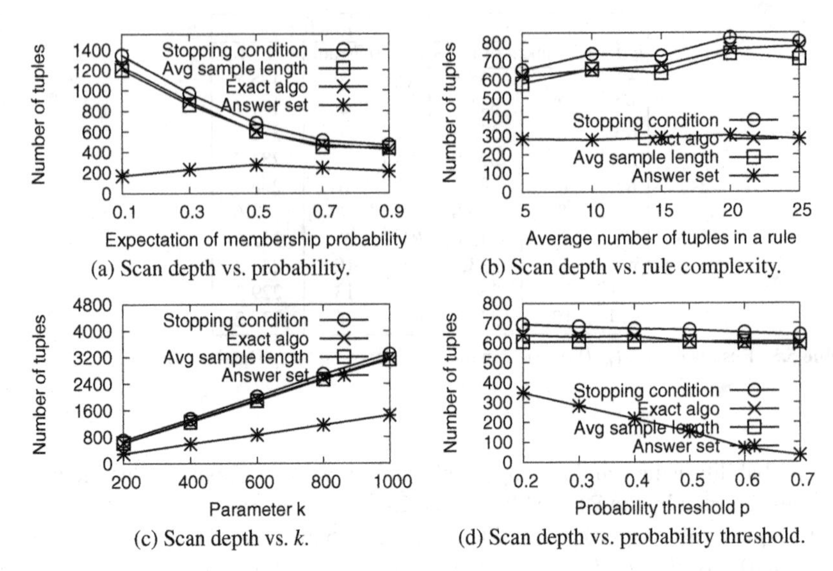

Fig. 5.4 Scan depth (each test data set contains 20,000 tuples and 2,000 generation rules).

By varying p in a top-(p,l) query, we can see the tradeoff between the confidence and the highest ranks a tuple can get with the confidence.

If $k = 10$ and $p = 0.4$, then a PT-k query returns $\{R1, R2, R3, R4, R5, R6, R9, R10, R11, R14, R15\}$. If p is increased to 0.7, then the answer set is $\{R1, R2, R3, R5, R6, R9, R10, R11\}$.

5.6.2 Results on Synthetic Data Sets

To evaluate the query answering quality and the scalability of our algorithms, we generate various synthetic data sets. The membership probability values of independent tuples and multi-tuple generation rules follow the normal distribution $N(\mu_{P_t}, \sigma_{P_t})$ and $N(\mu_{P_R}, \sigma_{P_R})$, respectively. The rule complexity, i.e., the number of tuples involved in a rule, follows the normal distribution $N(\mu_{|R|}, \sigma_{|R|})$[2].

By default, a synthetic data set contains 20,000 tuples and 2,000 multi-tuple generation rules: 2,000 generation rules. The number of tuples involved in each multi-tuple generation rule follows the normal distribution $N(5,2)$. The probability values of independent tuples and multi-tuple generation rules follow the normal distribution $N(0.5, 0.2)$ and $N(0.7, 0.2)$, respectively. We test the probability threshold top-k queries with $k = 200$ and $p = 0.3$.

[2] The data generator is available at `http://www.cs.sfu.ca/~jpei/Software/PTKLib.rar`

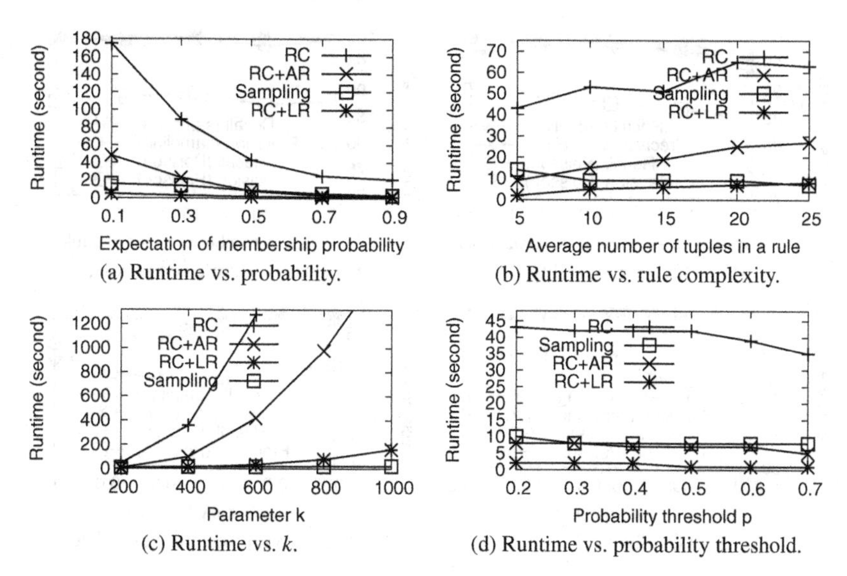

(a) Runtime vs. probability. (b) Runtime vs. rule complexity.

(c) Runtime vs. k. (d) Runtime vs. probability threshold.

Fig. 5.5 Efficiency (same settings as in Figure 5.4).

Since ranking queries are extensively supported by modern database management systems, we treat the generation of a ranked list of tuples as a black box, and test our algorithms on top of the ranked list.

First, to evaluate the efficient top-k probability computation techniques, we compare the exact algorithm, the sampling method, and the Poisson approximation based method for evaluating PT-k queries. The experimental results for top-(k,l) queries and top-(p,l) queries are similar to the results for PT-k queries. Therefore, we omit the details. For the exact algorithm, we compare three versions: Exact (using rule-tuple compression and pruning techniques only), Exact+AR (using aggressive reordering), and Exact+LR (using lazy reordering). The sampling method uses the two improvements described in Section 5.3. Then, we evaluate the online query answering techniques for PT-k queries, top-(k,l) queries and top-(p,l) queries.

5.6.2.1 Scan Depth

We test the number of tuples scanned by the methods (Figure 5.4). We count the number of distinct tuples read by the exact algorithm and the *sample length* as the average number of tuples read by the sampling algorithm to generate a sample unit. For reference, we also plot the number of tuples in the answer set, i.e., the tuples satisfying the probability threshold top-k queries, and the number of tuples computed by the general stopping condition discussed in Section 5.4.

In Figure 5.4(a), when the expected membership probability is high, the tuples at the beginning of the ranked list likely appear, which reduce the probabilities of

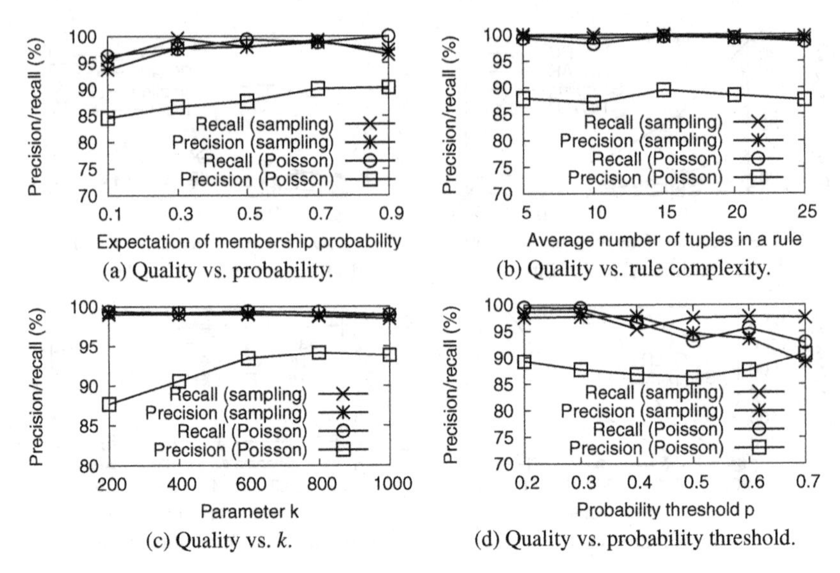

(a) Quality vs. probability.

(b) Quality vs. rule complexity.

(c) Quality vs. k.

(d) Quality vs. probability threshold.

Fig. 5.6 The approximation quality of the sampling method and the Poisson approximation-based method.

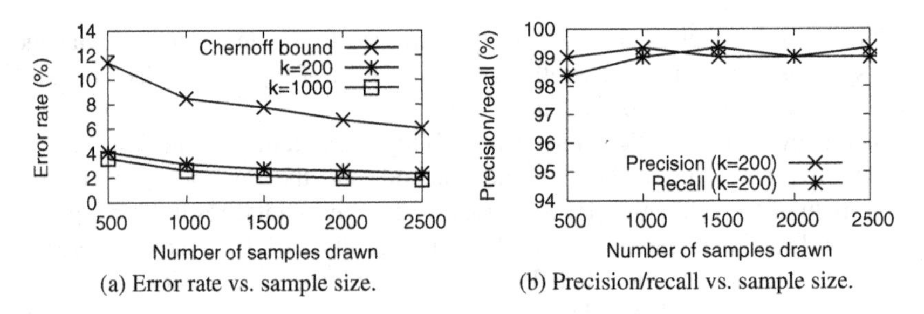

(a) Error rate vs. sample size.

(b) Precision/recall vs. sample size.

Fig. 5.7 The approximation quality of the sampling-based method.

the lower ranked tuples to be ranked in the top-k lists in possible worlds. If the membership probability of each tuple is very close to 1, then very likely we can prune all the tuples after the first k tuples are scanned. In contrary, if the expectation of the membership probability is low, then more tuples have a chance to be in the top-k lists of some possible worlds. Consequently, the methods have to check more tuples.

In Figure 5.4(b), when the rule complexity increases, more tuples are involved in a rule. The average membership probability of those tuples decreases, and thus more tuples need to be scanned to answer the query. In Figure 5.4(c), both the scan depth and the answer set size increase linearly when k increases, which is intuitive. In Figure 5.4(d), the size of the answer set decreases linearly as the probability threshold p increases. However, the number of tuples scanned decreases much slower. As

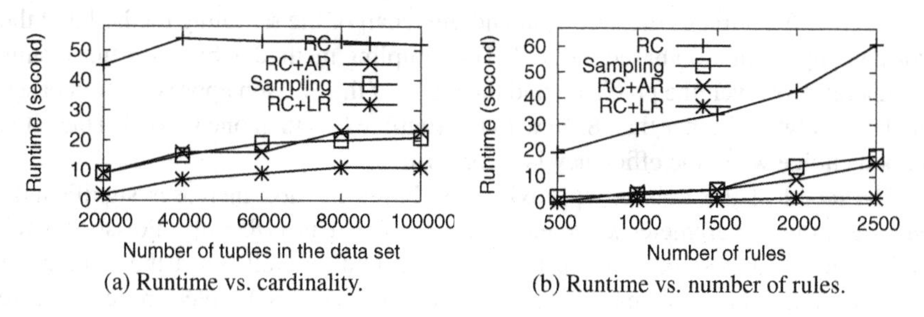

(a) Runtime vs. cardinality.

(b) Runtime vs. number of rules.

Fig. 5.8 Scalability.

discussed in Section 5.2.3, a tuple t failing the probability threshold still has to be retrieved if some tuples ranked lower than t may satisfy the threshold.

Figure 5.4 verifies the effectiveness of the pruning techniques discussed in Section 5.2.3. With the pruning techniques, the exact algorithm only accesses a small portion of the tuples in the data set. Interestingly, the average sample length is close to the number of tuples scanned in the exact algorithm, which verifies the effectiveness of our sampling techniques. Moreover, the exact algorithm and the sampling algorithm access fewer tuples than the number computed by the general stopping condition, while the number computed by the stopping condition is close to the real stopping point, which shows the effectiveness of the stopping condition.

5.6.2.2 Efficiency and Approximation Quality

Figure 5.5 compares the runtime of the three versions of the exact algorithm and the sampling algorithm with respect to the four aspects tested in Figure 5.4. The runtime of the Poisson approximation based method is always less than one second, so we omit it in Figure 5.5 for the sake of the readability of the figures. We also count the number of times in the three versions of the exact algorithm that subset probability values are computed. The trends are exactly the same as their runtime, therefore, we omit the figures here. The results confirm that the rule-tuple compression technique and the reordering techniques speed up the exact algorithm substantially. Lazy reordering always outperforms aggressive reordering substantially.

Compared to the exact algorithm, the sampling method is generally more stable in runtime. Interestingly, the exact algorithm (Exact+LR) and the sampling algorithm each has its edge. For example, when k is small, the exact algorithm is faster. The sampling method is the winner when k is large. As k increases, more tuples need to be scanned in the exact algorithm, and those tuples may be revisited in subset probability computation. But the only overhead in the sampling method is to scan more tuples when generating a sample unit, which is linear in k. This justifies the need for both the exact algorithm and the sampling algorithm.

Figure 5.6 compares the precision and the recall of the sampling method and the Poisson approximation based method. The sampling method achieves better results in general. However, the precision and the recall of the Poisson approximation based method is always higher than 85% with the runtime less than one second. Thus, it is a good choice when the efficiency is a concern.

The recall of the Poisson approximation based method increases significantly when the query parameter k increases. As indicated in [185], the Poisson distribution approximates the Poisson binomial distribution well when the number of Poisson trials is large. When the parameter k increases, more tuples are read before the stopping condition is satisfied. Thus, the Poisson approximation based method provides better approximation for the top-k probability values.

Figure 5.7(a) tests the average error rate of the top-k probability approximation using the sampling method. Suppose the top-k probability of tuple t is $Pr^k(t)$, and the top-k probability estimated by the sampling method is $\widehat{Pr}^k(t)$, the average error rate is defined as $\frac{\sum_{Pr^k(t)>p} |Pr^k(t)-\widehat{Pr}^k(t)|/Pr^k(t)}{|\{t|Pr^k(t)>p\}|}$. For reference, we also plot the error bound calculated from the Chernoff-Hoeffding bound [182] given the sample size. We can clearly see that the error rate of the sampling method in practice is much better than the theoretical upper bound.

Moreover, Figure 5.7(b) shows the precision and recall of the sampling method. The precision is the percentage of tuples returned by the sampling method that are in the actual top-k list returned by the exact algorithm. The recall is the percentage of tuples returned by the exact method that are also returned by the sampling method. The results show that the sampling method only needs to draw a small number of samples to achieve good precision and recall. With a larger k value, more samples have to be drawn to achieve the same quality

5.6.2.3 Scalability

Last, Figure 5.8 shows the scalability of the exact algorithm and the sampling algorithm. In Figure 5.8(a), we vary the number of tuples from $20,000$ to $100,000$, and set the number of multi-tuple rules to 10% of the number of tuples. We set $k = 200$ and $p = 0.3$. The runtime increases mildly when the database size increases. Due to the pruning rules and the improvement on extracting sample units, the scan depth (i.e., the number of tuples read) in the exact algorithm and the sampling algorithm mainly depends on k and is insensible to the total number of tuples in the data set.

In Figure 5.8(b), we fix the number of tuples to $20,000$, and vary the number of rules from 500 to $2,500$. The runtime of the algorithms increases since more rules lead to smaller tuple probabilities and more scans tuples back and forth in the span of rules. However, the reordering techniques can handle the rule complexity nicely, and make Exact+AR and Exact+LR scalable.

In all the above situations, the runtime of the Poisson approximation based method is insensitive to those factors, and remains within 1 second.

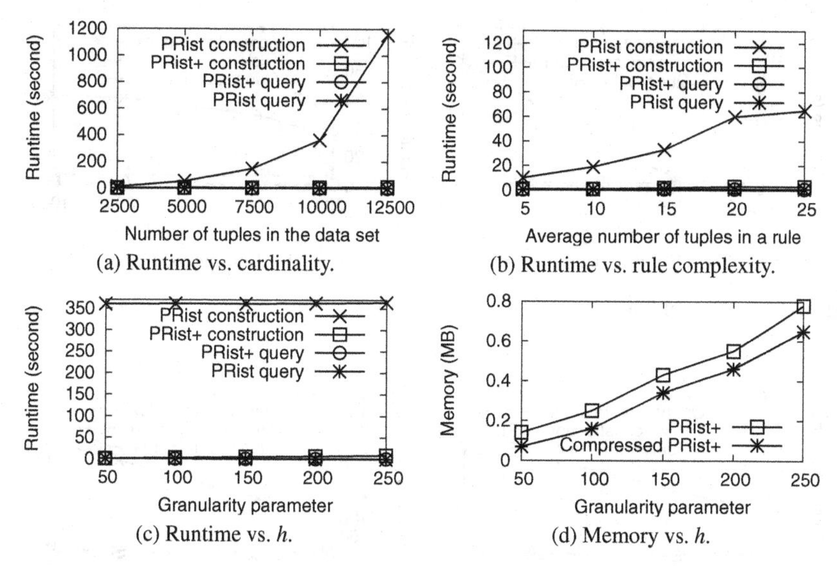

(a) Runtime vs. cardinality.

(b) Runtime vs. rule complexity.

(c) Runtime vs. h.

(d) Memory vs. h.

Fig. 5.9 The time and memory usage of *PRist* and *PRist+*.

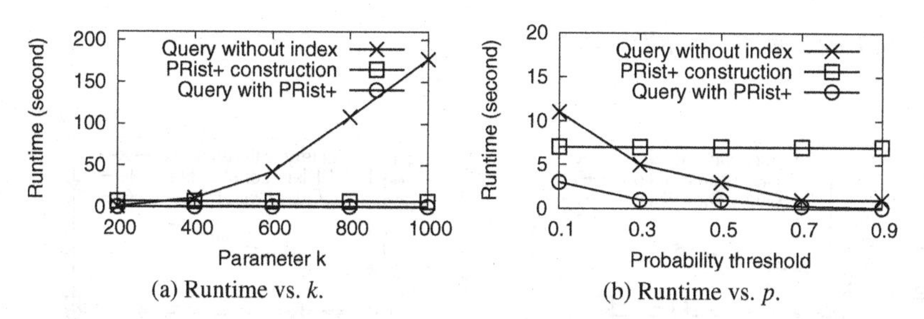

(a) Runtime vs. k.

(b) Runtime vs. p.

Fig. 5.10 Runtime of PT-k queries.

5.6.2.4 Online Query Answering

We evaluate the performance of the *PRist* and *PRist+* indices in answering PT-k queries, top-(k,l) queries, and top-(p,l) queries.

Figures 5.9(a)-(c) compare the construction time and average query answering time of *PRist* and *PRist+*. Clearly, *PRist+* can be constructed much more efficiently than *PRist* without sacrificing much efficiency in query answering.

The memory usage of the *PRist+* and compressed *PRist+* is shown in Figure 5.9(d). *PRist+* uses the compression techniques illustrated in Figure 5.3(b). The space to fully materialize the top-k probability for each tuple at each rank k is 8.3 MB. It shows that *PRist+* and *PRist* are much more space efficient than the full materialization method.

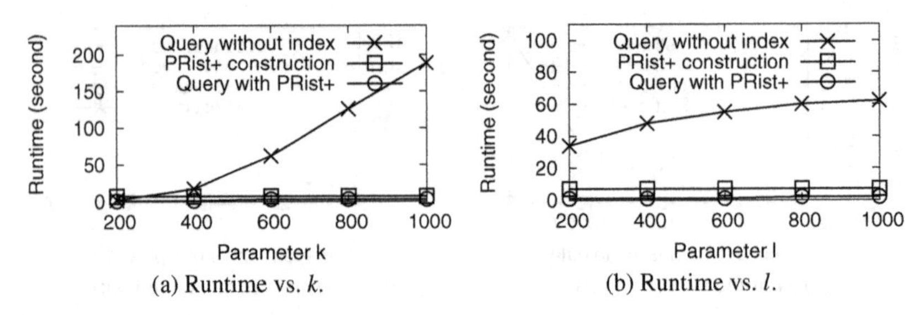

(a) Runtime vs. k. (b) Runtime vs. l.

Fig. 5.11 Runtime of top-(k,l) queries.

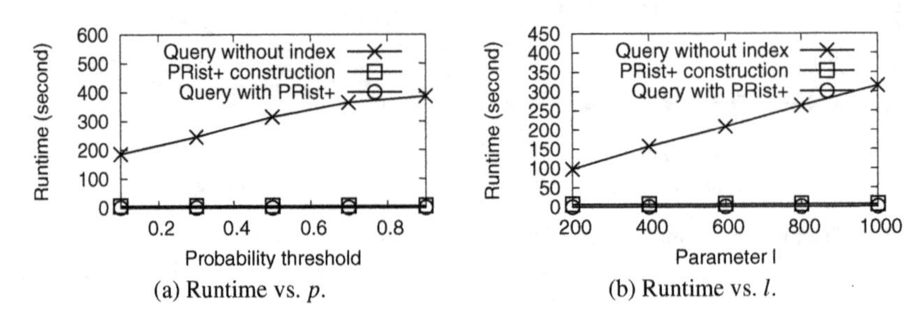

(a) Runtime vs. p. (b) Runtime vs. l.

Fig. 5.12 Runtime of top-(p,l) queries.

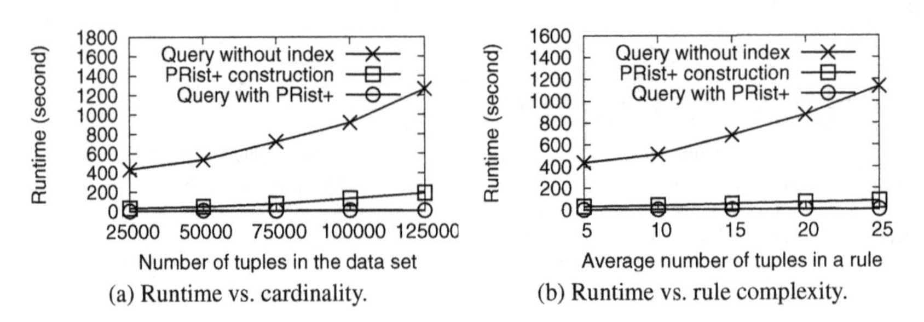

(a) Runtime vs. cardinality. (b) Runtime vs. rule complexity.

Fig. 5.13 Scalability.

We test the efficiency of the query evaluation methods. Since the query answering time based on *PRist* and *PRist+* is similar, here we only compare the efficiency of the query evaluation without index, and the efficiency of the query evaluation methods based on *PRist+*. PT-k queries, top-(k,l) queries and top-(p,l) queries are tested in Figures 5.10, 5.11 and 5.12, respectively. The construction time of *PRist+* is also plotted in those figures. There are 10,000 tuples and 1000 generation rules. The number of tuples in a rule follows the normal distribution $N(5,2)$. Clearly, the query evaluation methods based on *PRist+* have a dramatic advantage over the query answering methods without the index. Interestingly, even we construct a *PRist+*

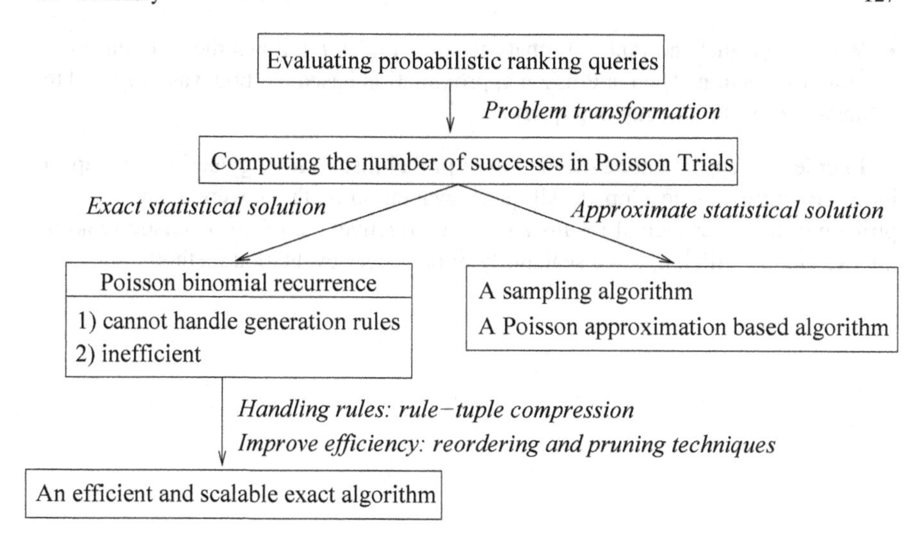

Fig. 5.14 The query evaluation algorithms for probabilistic ranking queries.

index on-the-fly to answer a query, in most cases it is still substantially faster than the query evaluation methods without indices.

Last, we test the scalability of the *PRist+* construction method and the query evaluation methods with respect to the number of tuples and the number of tuples in rules, respectively. Figure 5.13 shows that the methods are scalable and much more efficient than query answering without index.

5.7 Summary

In this chapter, we developed three methods, as shown in Figure 5.14, for answering probabilistic ranking queries defined in Section 2.2.2.

- We adopted the Poisson binomial recurrence [103] to compute the rank-k probabilities for independent tuples. Since the Poisson binomial recurrence cannot handle the tuples involved in generation rules, we developed a rule-tuple compression technique to transform the tuples in generation rules into a set of independent rule-tuples, so that the Poisson binomial recurrence can be applied. Moreover, to improve the efficiency, we devised two reordering techniques that reuse the computation. Last, we proposed several effective pruning techniques that reduce the number of tuples that we need to consider.
- We developed a sampling method to approximate the rank-k probabilities of tuples and compute the approximate answers to probabilistic ranking queries.

- We showed that the rank-k probability of a tuples t follows the Poisson binomial distribution. Then, a Poisson approximation based method was proposed to answer probabilistic ranking queries.

In order to support online evaluation of probabilistic ranking queries, a compact index structure was developed. All query evaluation methods were examined empirically. The experimental results show the effectiveness of probabilistic ranking queries and the efficiency and scalability of our query evaluation methods.

Chapter 6
Continuous Ranking Queries on Uncertain Streams

The uncertain data stream model developed in Section 2.3.1 characterizes the dynamic nature of uncertain data. Conceptually, an uncertain data stream contains a set of (potentially) infinite instances. To keep our discussion simple, we assume a synchronous model in this chapter. That is, at each time instant t $(t > 0)$, an instance is collected for an uncertain data stream. A sliding window W_{ω}^{t} selects the set of instances collected between time instants $t - \omega$ and t. The instances of each uncertain data stream in the sliding window can be considered as an uncertain object. We assume that the membership probabilities of all instances are identical. Some of our developed methods can also handle the case of different membership probabilities, which will be discussed in Section 6.5. A continuous probabilistic threshold top-k query reports, for each time instant t, the set of uncertain data streams whose top-k probabilities in the sliding window $W_{\omega}^{t}(\mathscr{O})$ are at least p.

In this chapter, we develop four algorithms systematically to answer a probabilistic threshold top-k query continuously: a deterministic exact algorithm, a randomized method, and their space-efficient versions using quantile summaries. An extensive empirical study using real data sets and synthetic data sets is reported to verify the effectiveness and the efficiency of our methods. Although we focus on monitoring probabilistic threshold top-k queries, the developed techniques can be easily extended to monitor other probabilistic ranking queries defined in Section 2.2.2 using the similar methods discussed in Chapter 5.

6.1 Exact Algorithms

In this section, we discuss deterministic algorithms to give exact answers to probabilistic threshold top-k queries. First, we extend the exact algorithm discussed in Section 5.2 in answering a probabilistic threshold top-k query in one sliding window. Then, we discuss how to share computation among overlapping sliding windows.

6.1.1 Top-k Probabilities in a Sliding Window

Consider a set of uncertain data streams $\mathscr{O} = \{O_1, \ldots, O_n\}$ and sliding window $W^t_\omega(\mathscr{O})$. $W^t_\omega(O_i) = \{O_i[t-\omega+1], \ldots, O_i[t]\}$ is the set of instances of O_i $(1 \leq i \leq n)$ in the sliding window. In this subsection, we consider how to rank the data streams according to their instances in sliding window $W^t_\omega(\mathscr{O})$. When it is clear from context, we write $W^t_\omega(\mathscr{O})$ and $W^t_\omega(O_i)$ simply as $W(\mathscr{O})$ and $W(O_i)$, respectively.

We reduce computing the top-k probability of a stream O into computing the top-k probabilities of instances of O.

Definition 6.1 (Top-k probability of instance). For an instance o and a top-k query Q^k, the *top-k probability* of o, denoted by $Pr^k(o)$, is the probability that o is ranked in the top-k lists in possible worlds. That is, $Pr^k(o) = \frac{\|\{w \in \mathscr{W} | o \in Q^k(w)\}\|}{\|\mathscr{W}\|}$. ∎

Following with Definitions 2.11 and 2.14, we have the following.

Corollary 6.1 (Top-k probability). *For an uncertain data stream O, a sliding window $W^t_\omega(\mathscr{O})$ and a top-k query Q^k,*

$$Pr^k(O) = \sum_{o \in W^t_\omega(O)} Pr^k(o) Pr(o) = \frac{1}{\omega} \sum_{o \in W^t_\omega(O)} Pr^k(o).$$

∎

We sort all instances according to their scores. Let R denote the ranking order of instances. For two instances o_1, o_2, we write $o_1 \prec o_2$ if o_1 is ranked before (i.e., better than) o_2 in R. Clearly, the rank of an instance o of stream O in the possible worlds depends on only the instances of other streams that are ranked better than o. We capture those instances as the dominant set of o.

Definition 6.2 (Dominant set). Given a set of streams $\mathscr{O}$, a sliding window W, and a top-k query Q^k, for an instance o of stream $O \in \mathscr{O}$, the **dominant set** of o is the set of instances of streams in $\mathscr{O} - \{O\}$ that are ranked better than o, denoted by $DS(o) = \{o' \in W(\mathscr{O} - O) | o' \prec o\}$. ∎

In a possible world w, an instance o is ranked the i-th place if and only if there are $(i-1)$ instances in $DS(o)$ appearing in w, and each of those instances is from a unique stream.

Based on this observation, for instance o and stream O' such that $o \notin O'$, we denote by $O' \prec o$ in a possible world w if there exists $o' \in W(O')$, $o' \prec o$, and o' and o appear in w. Apparently, we have

$$Pr(O' \prec o) = \sum_{o' \in DS(o), o' \in O'} Pr(o') Pr(o) = \frac{1}{\omega^2} \|DS(o) \cap W(O')\|.$$

Let $Pr(DS(o), i)$ be the probability that i instances in $DS(o)$ from unique streams appear in a possible world. Then, the top-k probability of o can be written as

$$Pr^k(o) = Pr(o) \sum_{i=0}^{k-1} Pr(DS(o), i) = \frac{1}{\omega} \sum_{i=0}^{k-1} Pr(DS(o), i).$$

For an instance o, since the events $O' \prec o$ for $O' \in \mathcal{O} - \{O\}$ are independent, we can view $DS(o)$ as a set of independent random binary trials, where each trial $X_{O'}$ is corresponding to an uncertain object O', $Pr(X_{O'} = 1) = Pr(O' \prec o)$, and $Pr(X_{O'} = 1) = 1 - Pr(X_{O'} = 1)$. The event that a trial takes value 1 is called a success. Since the probability that each trial takes value 1 is not identical, the total number of successes in $DS(o)$ follows the Poisson binomial distribution [103]. Thus, $Pr(DS(o), i)$ can be computed using the Poisson binomial recurrence given in Theorem 5.2.

The cost of sorting all instances in a sliding window is $O(n\omega \log(n\omega))$. To compute the top-k probability of each instance, the Poisson binomial recurrence is run and takes cost $O(kn)$ in time. Since there are $n\omega$ instances in the sliding window, the overall time complexity is $O(kn^2\omega + n\omega\log(n\omega))$.

Example 6.1 (Poisson binomial recurrence). Table 2.3 shows 4 *uncertain streams A, B, C, and D. For each instance, a ranking score is given. The ranking order is the ranking score descending order: the larger the ranking score, the better the instance is ranked.*

Let us consider the sliding window W_3^t *(i.e., the first three columns of instances in the figure), and compute the top-2 probability of* c_2. *The dominant set is* $DS(c_2) = \{a_1, a_2, a_3, d_3\}$. *Thus,* $p_1 = Pr(A \prec c_2) = Pr(a_1) + Pr(a_2) + Pr(a_3) = 1$, $p_2 = Pr(B \prec c_2) = 0$, *and* $p_3 = Pr(D \prec c_2) = Pr(d_3) = \frac{1}{3}$.

Using Theorem 5.2, let $S_1 = \{A\}$, $S_2 = \{A, B\}$ *and* $S_3 = \{A, B, D\}$. *For* S_1, *we have* $Pr(S_1, 0) = 1 - p_1 = 0$ *and* $Pr(S_1, 1) = p_1 = 1$.

For S_2, *we have* $Pr(S_2, 0) = (1 - p_2)Pr(S_1, 0) = 0$ *and* $Pr(S_2, 1) = p_2 Pr(S_1, 0) + (1 - p_2)Pr(S_1, 1) = 1$.

For S_3, *we have* $Pr(S_3, 0) = (1 - p_3)Pr(S_2, 0) = 0$ *and* $Pr(S_3, 1) = p_3 Pr(S_2, 0) + (1 - p_3)Pr(S_2, 1) = \frac{2}{3}$.

Thus, $Pr^2(c_2) = Pr(c_2)(Pr(S_3, 0) + Pr(S_3, 1)) = \frac{2}{9}$. ■

If we sort all the instances in sliding window $W(\mathcal{O})$ in the ranking order, then by one scan of the sliding window we can calculate the top-k probabilities for all instance. For each stream O, we only need to keep the following two pieces of information during the scan. First, we keep the number of instances in O that have been scanned. Suppose there are l such instances, then the probability of O in the Poisson recurrence is $\frac{l}{\omega}$. Second, we maintain the sum of the top-k probabilities of those scanned instances of O.

In practice, when a top-k query is raised, $k \ll n$ often holds where n is the total number of streams. In such a case, some streams can be pruned in the computation.

Theorem 6.1 (Pruning instances in a stream). *For an uncertain stream O, a top-k query* Q_p^k *with probability threshold p, and all instances in* $W(O)$ *sorted in the ranking order* $o_1 \prec \cdots \prec o_\omega$, *if there exists* i $(1 \le i \le \omega)$ *such that*

$$Pr^k(o_i) < \frac{p - \sum_{j=1}^{i-1} Pr^k(o_j)}{\omega - i + 1} \tag{6.1}$$

then $Pr^k(O) < p$.

Moreover, $Pr^k(O) \geq p$ *if there exists* i $(1 \leq i \leq \omega)$ *such that* $\sum_{j=1}^{i-1} Pr^k(o_j) \geq p$.

Proof. The first part: If there exists i such that $Pr^k(o_i) < \frac{p-\sum_{j=1}^{i-1} Pr^k(o_j)}{\omega-i+1}$, then

$$(\omega - i + 1)Pr^k(o_i) < p - \sum_{j=1}^{i-1} Pr^k(o_j).$$

Apparently, $Pr^k(o_1) \geq \cdots \geq Pr^k(o_\omega)$. Thus, we have

$$(\omega - i + 1)Pr^k(o_i) \geq \sum_{j=i}^{\omega} Pr^k(o_j).$$

Combining the above two inequalities, we have

$$\sum_{j=i}^{\omega} Pr^k(o_j) < p - \sum_{j=1}^{i-1} Pr^k(o_j).$$

Thus, $Pr^k(O) = \sum_{j=1}^{i-1} Pr^k(o_j) + \sum_{j=i}^{\omega} Pr^k(o_j) < p$.

To prove the second part, we only need to notice $Pr^k(O) = \sum_{j=1}^{\omega} Pr^k(o_j) \geq \sum_{j=1}^{i} Pr^k(o_j) \geq p$. ■

To use Theorem 6.1, for each stream O, if the last scanned instance in O satisfies one of the conditions in the theorem, the top-k probabilities of the remaining instances of O do not need to be computed.

For an object uncertain stream O whose top-k probability in sliding window W is smaller than the threshold p, we can derive the maximum number of instances scanned according to Theorem 6.1 as follows.

Corollary 6.2 (Maximum number of scanned instances). *For an uncertain stream* O*, a top-k query* Q_p^k *with probability threshold* p*, and all instances in* $W(O)$ *sorted in the ranking order* $o_1 \prec \cdots \prec o_\omega$*, the maximum number of instances scanned according to Theorem 6.1 is* $\lceil \frac{Pr^k(O)}{p} \omega \rceil + 1$.

Proof. Let o_t $(t > 1)$ be the first instance in $W(O)$ that satisfy Inequality 6.1. Then, o_i $(1 \leq i < t)$ does not satisfy Inequality 6.1. That is,

$$Pr^k(o_1) \geq \frac{p}{\omega}, \text{ and } Pr^k(o_i) \geq \frac{p - \sum_{j=1}^{i-1} Pr^k(o_j)}{\omega - i + 1} \text{ for } 1 < i < t$$

By induction, it is easy to show that for $1 \leq i < t$

$$\sum_{j=1}^{i} Pr^k(o_j) \geq i \times \frac{p}{\omega}$$

Since $Pr^k(O) = \sum_{j=1}^{t-1} Pr^k(o_j) + \sum_{m=t}^{\omega} Pr^k(o_m)$, we have

$$\sum_{j=1}^{t-1} Pr^k(o_j) = Pr^k(O) - \sum_{m=t}^{\omega} Pr^k(o_m) \geq (t-1) \times \frac{p}{\omega}$$

Therefore, $t \leq \lceil \frac{Pr^k(O)}{p} \omega \rceil + 1$. ■

Our second pruning rule is based on the following observation.

Lemma 6.1 (Sum of top-k probabilities). *For a set of uncertain data streams* $\mathcal{O}$*, a top-k query* Q^k*, and a sliding window* $W_\omega^t(\mathcal{O})$*,* $\sum_{O \in \mathcal{O}} Pr^k(O) = k$.

Proof. Using the definition of top-k probability, we have

$$\sum_{O \in \mathcal{O}} Pr^k(O) = \sum_{w \in \mathcal{W}, o \in Q^k(w)} Pr(w)$$

In a possible world w, $\sum_{o \in Q^k(w)} Pr(w) = k \cdot Pr(w)$. Using Corollary 2.4, we have $\sum_{O \in \mathcal{O}} Pr^k(O) = k \sum_{w \in \mathcal{W}} Pr(w) = k$. ■

Theorem 6.2 (Pruning by top-k probability sum). *Consider a set of uncertain data streams $\mathcal{O}$, a top-k query Q_p^k with probability threshold p, and a sliding window $W_\omega^t(\mathcal{O})$. Assume all instances in $W_\omega^t(\mathcal{O})$ are scanned in the ranking order, and $S \subset W_\omega^t(\mathcal{O})$ is the set of instances that are scanned. For a stream $O \in \mathcal{O}$, $Pr^k(O) < p$ if*

$$\sum_{o \in O \cap S} Pr^k(o) < p - (k - \sum_{o' \in S} Pr^k(o')).$$

Proof. Following with Lemma 6.1, we have

$$\sum_{o \in W_\omega^t(\mathcal{O}) - S} Pr^k(o) = k - \sum_{o' \in S} Pr^k(o').$$

If $\sum_{o \in O \cap S} Pr^k(o) < p - (k - \sum_{o' \in S} Pr^k(o'))$, then

$$\begin{aligned} Pr^k(O) &= \textstyle\sum_{o \in O} Pr^k(o) \\ &= \textstyle\sum_{o \in O \cap S} Pr^k(o) + \sum_{o \in O \cap (W_\omega^t(\mathcal{O}) - S)} Pr^k(o) \\ &< k - \textstyle\sum_{o' \in S} Pr^k(o') + p - (k - \sum_{o' \in S} Pr^k(o')) \\ &= p \end{aligned}$$

■

In summary, by sorting the instances in a sliding window in the ranking order and scanning the sorted list once, we can compute the top-k probability for each stream, and thus the exact answer to the top-k query on the window can be derived. The two pruning rules can be used to prune the instances and the streams.

6.1.2 Sharing between Sliding Windows

Using the method described in Section 6.1.1, we can compute the exact answer to a top-k query Q^k in one sliding window W^t. In the next time instant $(t+1)$, can we reuse some of the results in window W^t to compute the answer to Q^k in window W^{t+1}?

In this subsection, we first observe the compatible dominant set property, and then we explore sharing in computing answers to a top-k query on two consecutive sliding windows.

6.1.2.1 Compatible Dominant Sets

For an instance $o \in O$ that is in a window W^t, the top-k probability of o depends on only the number of instances from streams other than O that precede o in the ranking order. The ordering among those instances does not matter. Therefore, for an instance $o \in W^{t+1}$, if we can identify an instance o' in either W^t or W^{t+1} such that o and o' are compatible in terms of number of other preceding instances, then we can derive the top-k probability of o using that of o' directly. Technically, we introduce the concept of compatible dominant sets.

Definition 6.3 (Compatible dominant sets). Let $o \in O$ be an instance that is in window W^{t+1} and $DS^{t+1}(o)$ be the dominant set of o in W^{t+1}. For an instance $o_1 \in O$ and dominant set $DS(o_1)$, if for any stream $O' \neq O$, the number of instances from O' in $DS^{t+1}(o)$ and that in $DS(o)$ are the same, $DS^{t+1}(o)$ and $DS(o_1)$ are called **compatible dominant sets**. Please note that o may be the same instance as o_1, and $DS(o_1)$ can be in W^t or W^{t+1}. We consider $DS^{t+1}(o)$ and itself trivial compatible dominant sets. ■

Following with the Poisson binomial recurrence (Theorem 5.2), we immediately have the following result.

Theorem 6.3 (Compatible dominant sets). *If $DS^{t+1}(o)$ and $DS(o_1)$ are compatible dominant sets, for any $j \geq 0$, $Pr(DS^{t+1}(o), j) = Pr(DS(o_1), j)$ and $Pr^k(o) = Pr^k(o_1)$.*

Proof. If $DS^{t+1}(o)$ and $DS(o_1)$ are compatible dominant sets, then for any stream O' $(o, o_1 \notin O')$, $Pr[O' \in DS^{t+1}(o)] = Pr[O' \in DS(o_1)]$. Thus, following with Theorem 5.2, we have $\sum_{i=0}^{k-1} Pr(DS^{t+1}(o), i) = \sum_{j=0}^{k-1} Pr(DS(o_1), j)$. Therefore, $Pr^k(o) = Pr^k(o_1)$. ■

Compatible dominant sets can be employed directly to reduce the computation in window W^{t+1} using the results in window W^t and those already computed in window W^{t+1}. For any instance o, if the dominant set of o in W^{t+1} is compatible to some dominant set of o_1, then the top-k probability of o in W^{t+1} is the same as o_1. No recurrence computation is needed for o in W^{t+1}.

When the data streams evolve slowly, the instances from a stream may have a good chance to be ranked in the compatible places. Using compatible dominant sets can capture such instances and save computation.

Now, the problem becomes how to find compatible dominant sets quickly. Here, we give a fast algorithm which can be integrated to the top-k probability computation.

For each sliding window $W^t(\mathcal{O})$, we maintain the sorted list of instances in the window. When the window slides, we update the sorted list in two steps. First, we insert the new instances into the sorted list, but still keep the expired instances. We call the sorted list after the insertions the *expanded sorted list.*

We use an n-digit bitmap counter $c[1], \ldots, c[n]$, where n is the number of streams. At the beginning, $c[i] = 0$ for $1 \leq i \leq n$. We scan the expanded sorted list in the

Ranked list of the instances at time t and $t+1$
$a_2, a_1, c_1, a_3, a_4, d_3, c_4, c_2, d_2, b_1, b_4, b_2, d_1, d_4, c_3, b_3$

(a) The expanded sorted list.

Instance	Counter=[A,B,C,D]
a_2	[0, 0, 0, 0]
a_1	[1, 0, 0, 0]
c_1	[1, 0, 1, 0]
a_3	[1, 0, 1, 0]
a_4	[0, 0, 1, 0]
d_3	[0, 0, 1, 0]
c_4	[0, 0, 0, 0]
c_2	[0, 0, 0, 0]
d_2	[0, 0, 0, 0]
b_1	[0, 1, 0, 0]
b_4	[0, 0, 0, 0]
b_2	[0, 0, 0, 0]
d_1	[0, 0, 0, 1]
d_4	[0, 0, 0, 0]
c_3	[0, 0, 0, 0]
b_3	[0, 0, 0, 0]

(b) The bitmap counters.

Fig. 6.1 The sorted lists of instances in $SW(t-1)$ and $SW(t)$.

ranking order. If an expired instance or a new instance $o \in O_i$ is met, we set $c[i] = c[i] \oplus 1$.

For an instance $o \in O_i$ in the expanded list such that o is in both W^t and W^{t+1}, if all the bitmap counters, except for $c[i]$, are 0 right before o is read, then, for every instance $o' \in O_j$ $(i \neq j)$, $o' \prec o$ in the expanded sorted list, one of the following three cases may happen: (1) o' appears in both W^t and W^{t+1}; (2) $o' = O'_j[t-\omega+1]$ (i.e., o' appears in W^t only) and the new instance $O_j[t+1] \prec o$; or (3) $o' = O'_j[t+1]$ (i.e., appears in W^{t+1} only) and the expired instance $O'_j[t-\omega+1] \prec o$. In all the three cases, $DS^t(o)$ and $DS^{t+1}(o)$ are compatible if o does not arrive at time $t+1$.

If o arrives at time $t+1$, then we check the left and the right neighbors of o in the expanded sorted list. If one of them o' is from the same stream as o, then $DS(o)$ and $DS(o')$ are compatible.

We conduct Poisson recurrence for only instances which are in $W^{t+1}(\mathscr{O})$ and do not have a compatible dominance set. Otherwise, they are expired instances or their top-k probabilities can be obtained from the compatible dominant sets immediately. After one scan of the expanded sorted list, we identify all compatible dominant sets and also compute the top-k probabilities. Then, we remove from the expanded sorted list those expired instances. The current sliding window is processed. We are ready to slide the window to the next time instant $(t+2)$.

Example 6.2 (Compatible dominant set). Figure 6.1(a) shows the expanded sorted list of instances in sliding windows W_3^t and W_3^{t+1} in Table 2.3. At time $t+1$, the instances a_1, b_1, c_1, d_1 expire, and new instances a_4, b_4, c_4, d_4 arrive.

In Figure 6.1(b), we show the values of the bitmap counters during the scan of the expanded sorted list. Each instance in W_3^{t+1}, except for d_3, can find a compatible dominant set. We only need to conduct the Poisson recurrence computation of d_3 in W_3^{t+1}. ∎

6.1.2.2 Pruning Using the Highest Possible Rank

Consider an instance o in a sliding window W^t. As the window slides towards future, new instances arrive and old instances expire. As a result, the rank of o in the sliding windows may go up or down.

However, the instances arriving later than o or at the same time as o would never expire before o. In other words, the possible rank of o in the future sliding windows is bounded by those instances "no older" than o.

Lemma 6.2 (Highest possible rank). *For an instance $O[i]$ arriving at time i, in a sliding window $W_\omega^t(\mathcal{O})$ such that $t-\omega+1 < i \leq t$, let $\mathcal{R}_{O[i]} = \{O'[j] | O' \in \mathcal{O}, O' \neq O, j \geq i\}$. In any sliding window $W_\omega^{t'}$ such that $t' > t$, the rank of $O[i]$ cannot be less than $\|\mathcal{R}_{O[i]}\| + 1$.* ∎

Example 6.3 (Highest possible rank). Consider again the uncertain streams in Table 2.3. In window W_3^t, the rank of c_2 is 6. Among the 8 instances with time-stamp $t-1$ and t, there are 3 instances ranked better than c_2. The highest possible rank of c_2 in the future windows is 4. In window W^{t+1}, there are 5 instances arriving no earlier than c_2 and ranked better than c_2. The highest possible rank of c_2 in the future windows is 6. ∎

The highest possible rank of o can be used to derive an upper bound of the top-k probability of o in the future sliding windows.

Theorem 6.4 (Highest possible top-k probability). *For an instance o in sliding window W_ω^t with the highest possible rank $r \geq k\omega$, let $\rho = \frac{r-1}{\omega(n-1)}$, where n is the number of streams, in any window $W_\omega^{t'}$ ($t' \geq t$),*

$$Pr^k(o) \leq \frac{1}{\omega} \sum_{j=0}^{k-1} \binom{n}{j} \rho^j (1-\rho)^{n-j}$$

Proof. Given o with the highest possible rank r, there are $r-1$ instances from other objects ranked better than o. The sum of membership probabilities of those instances is $\frac{r-1}{\omega}$. The theorem follows with Lemma 5.15 directly. ∎

Corollary 6.3 (Pruning using highest possible rank). *For any instance $o \in O$, if $\sum_{o \in O} p_o < p$ and there exists p_o such that $Pr^k(o) \leq p_o$, then $Pr^k(O) < p$.* ∎

Algorithm 6.1 The exact algorithm.

Input: a set of uncertain streams $\mathcal{O}$, a sliding window width ω, a time instant $t+1$, a top-k query Q_p^k, and a probability threshold p
Output: Answer to top-k query in window W_ω^{t+1}
Method:
1: insert the new instances arriving at time $t+1$ into the sorted list in window W_ω^t;
2: initialize $Pr^k(O_i) = 0$ for each stream $O_i \in \mathcal{O}$;
3: compute the highest possible top-k probability for each stream $O_i \in \mathcal{O}$, remove O_i from consideration if it fails the probability threshold;
4: set counter $C[i] = 0$ for each stream $O_i \in \mathcal{O}$;
5: **for all** o in the expanded sorted list **do**
6: update the corresponding counter if o is expired or new;
7: compute $DS(o)$;
8: **if** $DS(o)$ has a compatible dominant set **then**
9: obtain $Pr^k(o)$ directly;
10: **else**
11: compute the probabilities $Pr(DS(I), j)$ $(0 \leq j < k)$;
12: **end if**
13: $Pr^k(O) = Pr^k(O) + Pr^k(o)$;
14: **if** $Pr^k(O) \geq p$ **then**
15: output O;
16: **end if**
17: check whether o can be used to prune some unread instances;
18: **if** all objects either are output or fail the probability threshold **then**
19: exit;
20: **end if**
21: **end for**

We need $O(1)$ space to maintain the highest possible rank for an instance. The overall space consumption is $O(n\omega)$ for a sliding window. Each time when new instances arrive, the highest possible ranks of all old instances are updated. The highest possible top-k probability of each stream is updated accordingly. This can be integrated into the top-k probability computation. For a stream O, once the upper bound of $Pr^k(O)$ fails the threshold, all instances in O do not need to be checked in the current window.

Algorithm 6.1 shows the complete exact algorithm. Compatible dominant sets can help to reduce the computation cost, however, although it works well in practice, in the worst case, the new instances may be ranked far away from the expired instances of the same stream, and thus no compatible dominant sets can be found. Thus, the time complexity of processing a sliding window, except for the first one, is $O(kn^2\omega + n\log(n\omega))$, where $O(n\log(n\omega))$ is the cost to insert the n new instances into the sorted list.

6.2 A Sampling Method

In this section, we propose a sampling method to estimate the top-k probability of each stream with a probabilistic quality guarantee.

For a stream O in a sliding window $W^t(\mathcal{O})$, we are interested in the event that O is ranked top-k. Let Z_O be the indicator to the event: $Z_O = 1$ if O is ranked top-k in $W^t(\mathcal{O})$; $Z_O = 0$ otherwise. Then, $Pr(Z_O = 1) = Pr^k(O)$.

To approximate the probability $Pr(Z_O = 1)$, we design a statistic experiment as follows. We draw samples of possible worlds and compute the top-k lists on the samples. That is, in a sample, for each stream O we select an instance o in window $W^t(O)$. A sample is a possible world. Then, we sort all instances in the sample in the ranking order and find the top-k list.

We repeat the experiment m times independently. Let $Z_{O,i}$ be the value of Z_O at the i-th run. Then, $\widetilde{E}[Z_O] = \frac{1}{m}\sum_{i=1}^{m} Z_{O,i}$ is an estimation of $E[Z_O] = Pr(Z_O = 1)$.

By using a sufficiently large number of samples, we can obtain a good approximation of $Pr^k(Z_O)$ with high probability. The methods follows the idea of unrestricted random sampling (also known as simple random sampling with replacement) [34]. The following minimum sample size can be derived from the well known Chernoff-Hoffding bound [182].

Theorem 6.5 (Minimum sample size). *For any stream O, let $Z_{O,1},\ldots,Z_{O,m}$ be the values of Z_O in m independent experiments, and $\widetilde{E}[Z_O] = \frac{1}{m}\sum_{i=1}^{m} Z_{O,i}$. For any δ $(0 < \delta < 1)$, ξ $(\xi > 0)$, if $m \geq \frac{3\ln\frac{2}{\delta}}{\xi^2}$, then $\Pr\{|\widetilde{E}[Z_O] - E[Z_O]| > \xi\} \leq \delta$.* ■

For efficient implementation, we maintain an indicator variable for each stream. To avoid sorting instances repeatedly, we first sort all instances in a sliding window $W^t(\mathcal{O})$. When drawing a sample, we scan the sorted list from the beginning, and select an instance $o \in O$ in probability $\frac{1}{\omega}$ if stream O has no instance in a sample yet. If an instance is chosen, the corresponding stream indicator is set. When the sample already contains k instances, the scan stops since the instances sampled later cannot be in the top-k list. The sample can then be discarded since it will not be used later.

The space complexity of the sampling method is $O(n\omega)$, because all instances in the sliding window have to be stored. The time complexity is $O(mn\omega + n\omega\log(n\omega))$ for the first window and $O(mn\omega + n\log(n\omega))$ for other windows where m is the number of samples drawn, since the n new instances can be inserted into the sorted list in $W^t(\mathcal{O})$ to form the sorted list in $W^{t+1}(\mathcal{O})$.

6.3 Space Efficient Methods

In the exact algorithm and the sampling algorithm, we need to store the sliding window in main memory. In some applications, there can be a large number of streams and the window width is non-trivial. In this section, we develop the space efficient methods using approximate quantile summaries. The central idea is to use quantiles to summarize instances in streams. Since computing exact quantiles in data streams is costly, we seek for high quality approximation.

Both the exact algorithm in Section 6.1 and the sampling method in Section 6.2 can be applied on the approximate quantile summaries of uncertain data streams. Using quantiles is a trade-off between space requirement and query answering accuracy. The distribution of an object is represented in a higher granularity level using quantiles, and thus the query results are approximate. However, we show that using approximate quantiles can save substantial space in answering top-k queries on uncertain streams with high quality guarantees.

6.3.1 Top-k Probabilities and Quantiles

Definition 6.4 (Quantile). Let $o_1 \prec \cdots \prec o_\omega$ be the sorted list of instances in the ranking order in a sliding window $W^t_\omega(O)$. The **ϕ-quantile** ($0 < \phi \leq 1$) of $W^t_\omega(O)$ is instance $o_{\lceil \phi\omega \rceil}$. A **$\phi$-quantile summary** of $W^t_\omega(O)$ is o_1 and a list of instances $o_{i\lceil \phi\omega \rceil}$ ($1 \leq i \leq \lceil \frac{1}{\phi} \rceil$).

The ϕ-quantile summary of $W^t_\omega(O)$ partitions the instances of $W^t_\omega(O)$ into $\lceil \frac{1}{\phi} \rceil$ **intervals** (in the values of the ranking function), with $\lceil \phi\omega \rceil$ instances in each interval. The first interval $t_1 = [o_1, o_{\lceil \phi\omega \rceil}]$. Generally, the i-th ($1 < i \leq \lceil \frac{1}{\phi} \rceil$) interval $t_i = (o_{\lceil (i-1)\phi\omega \rceil}, o_{\lceil i\phi\omega \rceil}]$. Since the membership probability of each instance is $\frac{1}{\omega}$, the **membership probability** of each interval t_i is $Pr(t_i) = \frac{1}{\omega}\phi\omega = \phi$. ■

Example 6.4 (A quantile summary). Consider a window $W^t_9(O)$ where the sorted list of instance scores is (21, 20, 12, 10, 9, 5, 4, 3, 2). *Then,* 12, 5 *and* 2 *are the* $\frac{1}{3}$, $\frac{2}{3}$, *and* 1*-quantiles of $W^t_9(O)$, respectively. The $\frac{1}{3}$-quantile summary of O is* $(21, 12, 5, 2)$ *which partitions the instances into three intervals:* $t_1 = [21, 12]$, $t_2 = (12, 5]$, *and* $t_3 = (5, 2]$. *The membership probability of each interval is* $\frac{1}{3}$. ■

We can use quantiles to approximate the top-k probabilities of streams.

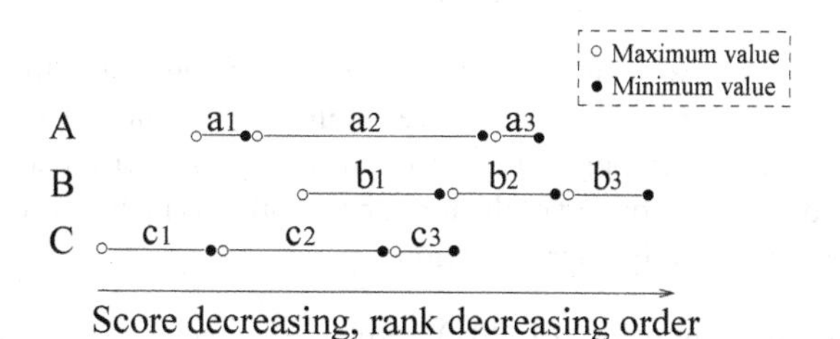

Fig. 6.2 The quantile summaries of three streams.

Example 6.5 (Approximating top-k probability). Consider three streams A, B, and C and their quantile summaries in window $W(\{A,B,C\})$, as shown in Figure 6.2, where a_i, b_i and c_i ($1 \leq i \leq 3$) are the intervals of $W(A)$, $W(B)$ and $W(C)$, respectively. The membership probability of each interval is $\frac{1}{3}$.

To compute the upper bound of the top-2 probability of instances falling into b_1, we let all instances in b_1 take the maximum value $b_1.MAX$. Moreover, since intervals a_2 and c_2 cover $b_1.MAX$, we let all instances in a_2 and c_2 take the minimum value $a_2.MIN$ and $c_2.MIN$, respectively. Thus, the probability that A is ranked better than b_1 is at least $Pr(a_1) = \frac{1}{3}$, and the probability that C is ranked better than b_1 is at least $Pr(c_1) = \frac{1}{3}$. The upper bound of the top-2 probability of b_1 is $Pr(b_1) \times (1 - \frac{1}{3} \times \frac{1}{3}) = \frac{8}{27}$.

Using the similar idea, we can verify that $\frac{1}{9}$ is the lower bound of the top-2 probability of b_1. ■

Using the idea illustrated in Example 6.5 and the Poisson binomial recurrence, we can have the general upper and lower bounds of the top-k probabilities of intervals.

Theorem 6.6 (Upper and lower bounds). *To answer a top-k query on sliding window $W(\mathcal{O})$, for an interval t in a ϕ-quantile summary of $W(O)$ ($O \in \mathcal{O}$),*

$$Pr^k(t) \leq Pr(t) \sum_{i=0}^{k-1} Pr(UDS(t), i)$$

and

$$Pr^k(t) \geq Pr(t) \sum_{i=0}^{k-1} Pr(LDS(t), i)$$

where $Pr^k(t) = \sum_{o \in W(O), o \in t} Pr^k(o)$, $UDS(t) = \{o' | o' \in t', t'$ is an interval in a ϕ-quantile summary of $W(O'), O' \neq O, t'.MIN \geq t.MAX\}$, and $LDS(t) = \{o' | o' \in t', t'$ is an interval in a ϕ-quantile summary of $W(O'), O' \neq O, t'.MAX \geq t.MIN\}$.

Proof. Since $UDS(t) \subseteq DS(t)$, $\sum_{i=1}^{k-1} Pr(UDS(t), i) \geq \sum_{j=1}^{k-1} Pr(DS(t), j)$. Similarly, $DS(t) \subseteq LDS(t)$, so we have $\sum_{j=1}^{k-1} Pr(DS(t), j) \geq \sum_{i=1}^{k-1} Pr(LDS(t), i)$. Moreover, since $Pr^k(t) = Pr(t) \sum_{j=1}^{k-1} Pr(DS(t), j)$, the conclusions hold. ■

Using Theorem 6.6, we can easily derive the upper bound and the lower bound of a stream by summing up the upper/lower bounds of all intervals of the quantile summary of the stream. Importantly and interestingly, the difference between the upper bound and the lower bound of the top-k probability is up to 2ϕ, which provides a strong quality guarantee in approximation.

Theorem 6.7 (Approximation quality). *For a stream O, let $\mathcal{U}(Pr^k(O))$ and $\mathcal{L}(Pr^k(O))$ be the upper bound and the lower bound of $Pr^k(O)$ derived from Theorem 6.6, respectively. $\mathcal{U}(Pr^k(O)) - \mathcal{L}(Pr^k(O)) \leq 2\phi$.*

Proof. To prove Theorem 6.7, we need the following lemma.

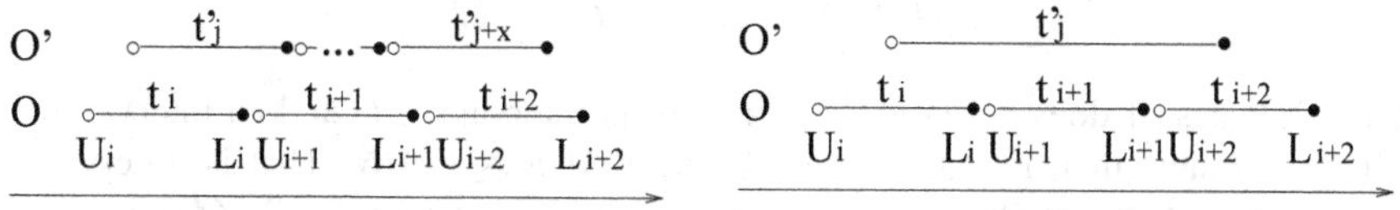

Fig. 6.3 Two cases in the proof of Theorem 6.7.

Lemma 6.3 (Monotonicity). *Let t_i, t_{i+1} and t_{i+2} be three consecutive intervals in the ϕ-quantile summary of $W(O)$. Let $\mathcal{L}(Pr^k(t_i))$ and $\mathcal{U}(Pr^k(t_{i+2}))$ be the lower bound of $Pr^k(t_i)$ and the upper bounds of $Pr^k(t_{i+2})$, respectively, derived from Theorem 6.6. If for any interval t' in the ϕ-quantile summary of O' ($O' \neq O$), $t'.MAX > t_i.MIN$, $t'.MIN > t_{i+2}.MAX$, then $\mathcal{L}(Pr^k(t_i)) \geq \mathcal{U}(Pr^k(t_{i+2}))$.*

Proof. For intervals t_i and t_{i+2}, we have $LDS(t_i) = \{t' | t' \in O', t'.MAX > t_i.MIN\}$, and $UDS(t_{i+2}) = \{t' | t' \in O', t'.MIN \geq t_{i+2}.MAX\}$. Using the assumption in the lemma, we have $LDS(t_i) \subset UDS(t_{i+2})$. Thus, $\mathcal{L}(Pr^k(t_i)) \geq \mathcal{U}(Pr^k(t_{i+2}))$.

We consider the following two cases:

Case 1. For any interval $t \in W(O)$ and $t' \in W(O')$ ($O' \neq O$), if $t'.MAX > t.MAX$, then $t'.MIN > t.MIN$, as illustrated in Figure 6.3(a). Lemma 6.3 holds in this case.

According to the definitions of $\mathcal{U}(Pr^k(O))$ and $\mathcal{L}(Pr^k(O))$, $\mathcal{U}(Pr^k(O)) = \sum_{i=1}^{\frac{1}{\phi}} \mathcal{U}(Pr^k(t_i))$ and $\mathcal{L}(Pr^k(O)) = \sum_{i=1}^{\frac{1}{\phi}} \mathcal{L}(Pr^k(t_i))$. Thus,

$$\begin{aligned} & \mathcal{U}(Pr^k(O)) - \mathcal{L}(Pr^k(O)) \\ = & \sum_{i=1}^{\frac{1}{\phi}-2} \left(\mathcal{U}(Pr^k(t_{i+2})) - \mathcal{L}(Pr^k(t_i))\right) + \mathcal{U}(Pr^k(t_1)) \\ & + \mathcal{U}(Pr^k(t_2)) - \mathcal{L}(Pr^k(t_{\frac{1}{\phi}-1})) - \mathcal{L}(Pr^k(t_{\frac{1}{\phi}})) \end{aligned} \tag{6.2}$$

Using Lemma 6.3, we have

$$\sum_{i=1}^{\frac{1}{\phi}-2} \left(\mathcal{U}(Pr^k(t_{i+2})) - \mathcal{L}(Pr^k(t_i))\right) < 0.$$

Thus, we have

$$\begin{aligned} & \mathcal{U}(Pr^k(O)) - \mathcal{L}(Pr^k(O)) \\ < & \mathcal{U}(Pr^k(t_1)) + \mathcal{U}(Pr^k(t_2)) - \mathcal{L}(Pr^k(t_{\frac{1}{\phi}-1})) - \mathcal{L}(Pr^k(t_{\frac{1}{\phi}})) \end{aligned} \tag{6.3}$$

Also, for $1 \leq i \leq \frac{1}{\phi}$, $Pr(t_i) = \phi$ and

$$0 \leq \mathcal{U}(Pr^k(t_i)) \leq Pr(t_i) \text{ and } 0 \leq \mathcal{L}(Pr^k(t_i)) \leq Pr(t_i).$$

Thus, we have

$$\mathcal{U}(Pr^k(t_1)) + \mathcal{U}(Pr^k(t_2)) - \mathcal{L}(Pr^k(t_{\frac{1}{\phi}-1})) - \mathcal{L}(Pr^k(t_{\frac{1}{\phi}})) \leq 2\phi \tag{6.4}$$

Plugging Inequalities 6.3 and 6.4 into Equation 6.2, we get

$$\mathscr{U}(Pr^k(O)) - \mathscr{L}(Pr^k(O)) \le 2\phi.$$

Case 2. If case 1 does not hold, i.e., there is an interval $t \in O$ and an interval $t' \in W(O) - O$ such that, $t'.MAX > t.MAX$ and $t'.MIN \le t.MIN$. That is, interval t' covers t completely, as illustrated in Figure 6.3(b). In that case, $\mathscr{U}(Pr(O' \prec t_{i+1})) - \mathscr{L}(Pr(O' \prec t_{i+1})) = Pr(t_j) = \phi$. Comparing to Case 1 where $\mathscr{U}(Pr(O' \prec t_{i+1})) - \mathscr{L}(Pr(O' \prec t_{i+1})) = Pr(t_j) + \ldots + Pr(t_{j+x}) = (x-1)\phi$, the difference between the upper bound and the lower bound is smaller. Therefore, in Case 2, $\mathscr{U}(Pr^k(O)) - \mathscr{L}(Pr^k(O))$ is even smaller than that in Case 1. The theorem holds. ■

For any object O, since $\mathscr{U}(Pr^k(O)) - \mathscr{L}(Pr^k(O)) \le 2\phi$, we can simply use $\frac{\mathscr{U}(Pr^k(O)) - \mathscr{L}(Pr^k(O))}{2}$ to approximate $Pr^k(O)$.

Corollary 6.4 (Approximation Quality). *For a stream $O \in W(\mathscr{O})$, let $\widetilde{Pr}^k(O) = \frac{\mathscr{U}(Pr^k(O)) - \mathscr{L}(Pr^k(O))}{2}$, then $\|\widetilde{Pr}^k(O) - Pr^k(O)\| \le \phi$.* ■

6.3.2 Approximate Quantile Summaries

Although using quantiles we can approximate top-k probabilities well, computing exact quantiles of streams by a constant number of scans still needs $\Omega(N^{\frac{1}{p}})$ space [187]. To reduce the cost in space, we use ε-approximate quantile summary which can still achieve good approximation quality.

Definition 6.5 (ε-approximate quantile). Let $o_1 \prec \cdots \prec o_\omega$ be the sorted list of instances in a sliding window $W(O)$. An **ε-approximate ϕ-quantile** $(0 < \phi \le 1)$ of $W(O)$ is an instance O_l where $l \in [\lceil(\phi - \varepsilon)\omega\rceil, \lceil(\phi + \varepsilon)\omega\rceil]$.

An **ε-approximate ϕ-quantile summary** of $W(O)$ is o_1 and a list of instances $o_{l_1}, \ldots, o_{l_{\lceil 1/\phi \rceil}}$, $l_i \in [\lceil(i\phi - \varepsilon)\omega\rceil, \lceil(i\phi + \varepsilon)\omega\rceil]$ $(1 \le i \le \lceil\frac{1}{\phi}\rceil)$.

The ε-approximate ϕ-quantile summary of $W(O)$ partitions the instances of $W(O)$ into $\lceil\frac{1}{\phi}\rceil$ **intervals**. The first interval $t_1 = [o_1, o_{l_1}]$, and generally the i-th $(1 < i \le \lceil\frac{1}{\phi}\rceil)$ interval $t_i = (q_{i-1}, q_i]$. ■

The number of instances in each interval is in $[(\phi - 2\varepsilon)\omega, (\phi + 2\varepsilon)\omega]$. Since the membership probability of each instance is $\frac{1}{\omega}$, the membership probability of each interval is within $[\phi - 2\varepsilon, \phi + 2\varepsilon]$.

Computing ε-Approximate quantiles in data streams is well studied [111, 112, 188, 189]. Both deterministic and randomized methods are proposed. In our implementation, we adopt the method of computing approximate quantile summaries in a sliding window proposed in [188], which is based on the GK-algorithm [112] that finds the approximate quantile over a data steam. The algorithm can continuously output the ε-approximate quantiles in a sliding window with space cost of $O(\frac{\log^2 \varepsilon\omega}{\varepsilon^2})$.

Then, how can we compute the upper bound and the lower bound of the top-k probability of a stream in a sliding window using its ε-approximate ϕ-quantile summaries?

Consider an interval $t_i = (o_{i-1}, o_i]$ in an ε-approximate ϕ-quantile summary. Suppose $R(o_{i-1})$ and $R(o_i)$ are the actual rank of o_{i-1} and o_i, respectively. Then, the actual number of instances in t_i is $R(o_i) - R(o_{i-1})$. The membership probability of t_i is $Pr(t_i) = \frac{R(o_i)-R(o_{i-1})}{\omega}$. However, since o_{i-1} and o_i are approximations of the $(i-1)\phi$- and $i\phi$-quantiles, respectively, their actual ranks are not calculated. Instead, we use $\widetilde{Pr}(t) = \phi$ to approximate the membership probability of t_i. Since $((i-1)\phi - \varepsilon)\omega \le R(o_{i-1}) \le ((i-1)\phi + \varepsilon)\omega$, and $(i\phi - \varepsilon)\omega \le R(o_i) \le (i\phi + \varepsilon)\omega$, we have $\|Pr(t) - \widetilde{Pr}(t)\| \le 2\varepsilon$.

Then, we compute the upper bound and the lower bound of $Pr^k(t)$, denoted by $\widetilde{\mathscr{U}}(Pr^k(t))$ and $\widetilde{\mathscr{L}}(Pr^k(t))$, respectively, using the approximate membership probability $\widetilde{Pr}(t)$, following with Theorem 6.6. In sequel, we can further derive the upper bound and the lower bound of a stream by summing up the upper bound and the lower bound of all intervals, respectively. The above approximation method has the following quality guarantee.

Theorem 6.8 (Approximation quality). *Given a stream O in a sliding window W, let $\widetilde{\mathscr{U}}(Pr^k(O))$ and $\widetilde{\mathscr{L}}(Pr^k(O))$ be the upper and lower bounds of $Pr^k(O)$ computed using the ε-approximate ϕ-quantile summary of $W(O)$, then,*

$$\left\|\widetilde{\mathscr{U}}(Pr^k(O)) - \mathscr{U}(Pr^k(O))\right\| \le \varepsilon \tag{6.5}$$

and

$$\left\|\widetilde{\mathscr{L}}(Pr^k(O)) - \mathscr{L}(Pr^k(O))\right\| \le \varepsilon \tag{6.6}$$

Proof. Consider an ε-approximate $i\phi$-quantile $o_i \in O$. To analyze the approximation error introduced by o_i, we first assume that other quantiles o_j $(1 \le j \le \frac{1}{\phi},\ j \ne i)$ are exact. Suppose the real rank of o_i is $R(o_i)$, according to the definition of ε-approximate quantile, we have $(i\phi - \varepsilon)\omega \le R(o_i) \le (i\phi + \varepsilon)\omega$.

$t_i = (o_{i-1}, o_i]$ and $t_{i+1} = (o_i, o_{i+1}]$ are two intervals partitioned by o_i. The approximate numbers of instances in t_i and t_{i+1} are both $\phi\omega$.

If $R(o_i) < \phi$, then the actual number of instances in t_i is $R(o_i) - (\phi - 1)\omega < \phi\omega$, and the actual number of instances in t_{i+1} is $(\phi + 1)\omega - R(o_i) > \phi\omega$. That is, there are $\phi\omega - R(o_i) \le \varepsilon\omega$ instances that are actually in t_{i+1}, but are counted into t_i due to the ε-approximate quantile o_i. Thus, the error introduced by o_i is at most $\|\frac{\varepsilon}{\phi}(\mathscr{U}(Pr^k(t_i)) - \mathscr{U}(Pr^k(t_{i+1})))\|$. Similarly, if $R(o_i) < \phi$, the error introduced by o_i is at most $\|\frac{\varepsilon}{\phi}(\mathscr{U}(Pr^k(t_{i+1})) - \mathscr{U}(Pr^k(t_i)))\|$.

Generally, the maximum overall approximation error introduced by ε-approximate quantiles $o_1, \ldots, o_{\frac{1}{\phi}-1}$ is

$$\left\|\widetilde{\mathscr{U}}(Pr^k(O)) - \mathscr{U}(Pr^k(O))\right\| = \Sigma_{i=1}^{\frac{1}{\phi}-1} \left\|\frac{\varepsilon}{\phi}(\mathscr{U}(Pr^k(t_i)) - \mathscr{U}(Pr^k(t_{i+1})))\right\|$$
$$\le \frac{\varepsilon}{\phi}(\mathscr{U}(Pr^k(t_1)) - \mathscr{U}(Pr^k(t_{\frac{1}{\phi}}))) \le \varepsilon$$

Inequality 6.6 can be shown similarly. ■

For a sliding window $W(O)$ of a stream, we use $\frac{\widetilde{\mathscr{U}}(Pr^k(O))-\widetilde{\mathscr{L}}(Pr^k(O))}{2}$ as an approximation of $Pr^k(O)$.

Theorem 6.9 (Approximation Quality). *For a stream O and sliding window $W(O)$, let $\widetilde{Pr}^k(O) = \frac{\widetilde{\mathscr{U}}(Pr^k(O))-\widetilde{\mathscr{L}}(Pr^k(O))}{2}$, then $\|\widetilde{Pr}^k(O) - Pr^k(O)\| \leq \phi + \varepsilon$.*

Proof. Following with Theorems 6.8 and 6.7, we have

$$\widetilde{\mathscr{U}}(Pr^k(O)) - \widetilde{\mathscr{L}}(Pr^k(O)) \leq \mathscr{U}(Pr^k(O)) - \mathscr{L}(Pr^k(O)) + 2\varepsilon \leq 2\phi + 2\varepsilon$$

Theorem 6.9 follows with the above inequality directly. ■

6.3.3 Space Efficient Algorithms using Quantiles

The deterministic algorithm discussed in Section 6.1 and the sampling algorithm proposed in Section 6.2 can both be extended using approximate quantile summaries. Due to the loss of information in approximate quantile summaries, the extension of the deterministic algorithm only provides approximate answers.

Using approximate quantile summaries, each stream in a sliding window is represented by $\lceil \frac{1}{\phi} \rceil$ intervals. The upper bound and the lower bound of the top-k probability of each interval can be computed using either the deterministic method or the sampling method.

To compute the upper bound and the lower bound using the deterministic method, we first sort the maximum values and the minimum values of all intervals in the ranking order. Then, by scanning the sorted list once, we can compute the approximate upper bound and the approximate lower bound of the top-k probability of each interval. For each stream O, we maintain the upper bound and the lower bound of the number of instances in $W(O)$ that have been scanned, and the upper bound and the lower bound of $Pr^k(O)$ so far.

The time complexity of query evaluation in a sliding window using the above extended algorithm is $O(\frac{kn^2}{\phi} + \frac{n}{\phi}\log(n\frac{1}{\phi}))$. The upper bound of approximation error is $\phi + \varepsilon$.

To compute the upper bound and the lower bound using the sampling method, we draw m sample units uniformly at random as described in Section 6.2. The difference is, each sample unit contains an interval from each stream. For each interval t of stream O, we define an indicator $X_{t_{\mathscr{U}}}$ to the event that $\|\{t'|t'$ is an interval of $O', O' \neq O, t'.MIN > t.MAX\}\| < k$, and an indicator $X_{t_{\mathscr{L}}}$ to the event that $\|\{t'|t'$ is an interval of $O', O' \neq O, t'.MAX > t.MIN\}\| < k$. The indicator is set to 1 if the event happens; otherwise, the indicator is set to 0. Then, $\mathscr{U}(Pr^k(t)) = E[X_{t_{\mathscr{U}}}]$, and $\mathscr{L}(Pr^k(t)) = E[X_{t_{\mathscr{L}}}]$. Suppose in sample unit s, the value of $X_{t_{\mathscr{U}}}$ is $X^s_{t_{\mathscr{U}}}$, then the expectation $E[X_{t_{\mathscr{U}}}]$ can be estimated by $\frac{1}{m}\sum_{i=1}^{m} X^{s_i}_{t_{\mathscr{U}}}$. Similarly, $E[X_{t_{\mathscr{L}}}]$ can be estimated by $\frac{1}{m}\sum_{i=1}^{m} X^{s_i}_{t_{\mathscr{L}}}$.

The time complexity of query evaluation in a sliding window using the above sampling method is $O(mn\frac{1}{\phi})$, where m is the number of samples. If $m \geq \frac{3\ln\frac{2}{\delta}}{\xi^2}$, ($0 < \delta < 1$, $\xi > 0$), then, the upper bound of approximation error is $\phi + \varepsilon + \xi$ with a probability at least $1 - \delta$.

In the above two extended algorithms using approximate quantile summaries, the space complexity of the algorithms is reduced from $O(n\omega)$ to $O(n\frac{\log^2 \varepsilon\omega}{\varepsilon^2})$, which is the space complexity of computing ε approximate quantiles.

6.4 Experimental Results

In this section, we report a systematic empirical study. All the experiments were conducted on a PC computer with a 3.0 GHz Pentium 4 CPU, 1.0 GB main memory, and a 160 GB hard disk, running the Microsoft Windows XP Professional Edition operating system. Our algorithms were implemented in Microsoft Visual Studio 2005.

6.4.1 Results on Real Data Sets

To illustrate the effectiveness of probabilistic threshold top-k queries over sliding windows in real applications, we use the seismic data collected from the wireless sensor network monitoring volcanic eruptions[1]. 16 sensors were deployed at Reventador, an active volcano in Ecuador. Each of the 16 sensors continuously sampled seismic data, and the last 60 seconds of data from each node was examined at the base station. To detect the eruption, it is interesting to continuously report the top-k monitoring locations with the highest seismic values in the last 60 seconds.

The seismic data reported by each sensor is treated as an uncertain stream, and each data record is an instance. We test probabilistic threshold top-k queries with different parameter values on the data set. Since the results demonstrate the similar patterns, we only report the answers to a probabilistic threshold top-k query with $k = 5$ and $p = 0.4$ in this book. We consider a sliding window width of 60 instances per stream. The answers to the query in 10 consecutive sliding windows are reported in Table 6.1. As comparison, we also compute the average value of each stream in each sliding window and report the top-5 streams with the highest average seismic values.

The answers to the probabilistic threshold top-k query listed in Table 6.1 reveal the following interesting patterns. First, the seismic values reported by sensors O_2 and O_{16} are consistently among the top-5 with high confidence in the 10 sliding windows. The rankings of seismic values in those locations are stable. Second, the seismic values reported by sensor O_6 is among the top-5 with high confidence in the

[1] `http://fiji.eecs.harvard.edu/Volcano`

Window ID	PTK query ($k=5$, $p=0.4$)	Top-5 query on average data
W_1	O_2, O_6, O_{16}	$O_2, O_4, O_6, O_{14}, O_{16}$
W_2	O_2, O_6, O_{16}	$O_2, O_4, O_6, O_{14}, O_{16}$
W_3	O_2, O_6, O_{16}	$O_2, O_4, O_6, O_{14}, O_{16}$
W_4	O_2, O_4, O_6, O_{16}	$O_2, O_4, O_6, O_{14}, O_{16}$
W_5	O_2, O_6, O_{16}	$O_2, O_4, O_6, O_{14}, O_{16}$
W_6	O_2, O_{16}	$O_2, O_4, O_6, O_{14}, O_{16}$
W_7	O_2, O_{16}	$O_2, O_4, O_6, O_{14}, O_{16}$
W_8	O_2, O_{16}	$O_2, O_4, O_6, O_{14}, O_{16}$
W_9	O_2, O_{16}	$O_2, O_4, O_6, O_{14}, O_{16}$
W_{10}	O_2, O_{16}	$O_2, O_4, O_6, O_{14}, O_{16}$

Table 6.1 The answers to a probabilistic threshold top-*k* query in 10 consecutive sliding windows ($\omega = 60$) and the answers to a traditional top-*k* query on average values.

first 5 sliding windows. The rankings of seismic values reported by sensor O_6 drop after sliding window W_5. Third, the seismic values reported by sensor O_4 is ranked among top-5 with high confidence only in sliding window W_4.

A traditional top-5 query on the average seismic values in each sliding window reports $\{O_2, O_4, O_6, O_{14}, O_{16}\}$ consistently in the 10 sliding windows. They do not reflect the above interesting patterns.

Simple example shows that continuous probabilistic threshold top-*k* queries on uncertain streams provide meaningful information which cannot be captured by the traditional top-*k* queries on aggregate data.

6.4.2 Synthetic Data Sets

In this performance study, we use various synthetic data sets to evaluate the efficiency of the algorithms and the query evaluation quality. By default, a data set contains 100 uncertain streams, and the sliding window width is set to $\omega = 200$. Thus, there are 20,000 instances in each sliding window. The data in a sliding window is already held in main memory. The scores of instances from one stream follow a normal distribution. The mean μ is randomly picked from a domain $[0, 1000]$, and the variance σ is randomly picked from $[0, 10]$. We add 10% noise by using 10σ as the variance. Moreover, the query parameter $k = 20$, and the probability threshold $p = 0.4$. The number of samples drawn in the sampling algorithm is 1,000. In quantile summaries, $\phi = 0.1$, and $\varepsilon = 0.02$. The reported results are the average values in 5 sliding windows.

We test the following four algorithms: the *deterministic exact* method (*Det*) in Section 6.1.2; the *sampling* method (*Sam*) in Section 6.2; the *extended deterministic* method using quantile summaries (*Det-Q*) and the *sampling* method using quantile summaries (*Sam-Q*) in Section 6.3.3.

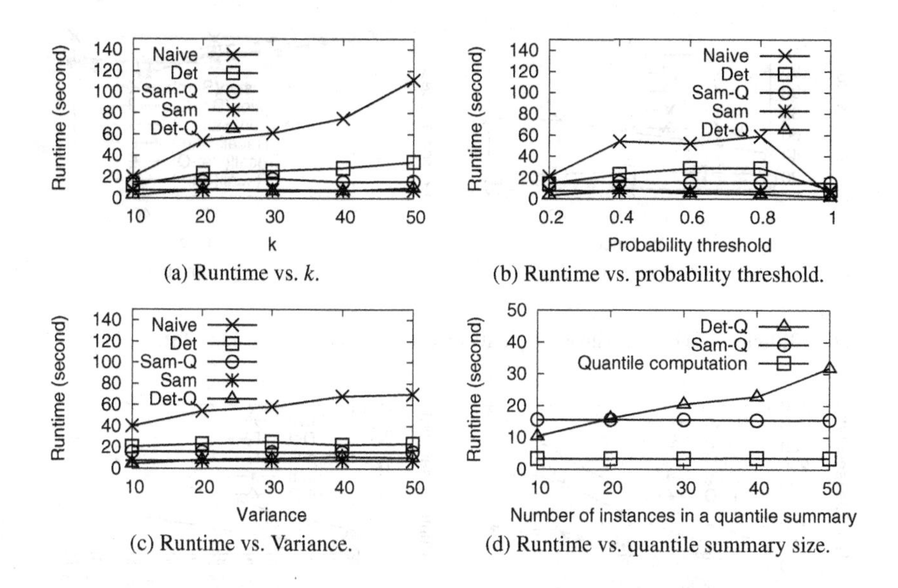

(a) Runtime vs. k. (b) Runtime vs. probability threshold.

(c) Runtime vs. Variance. (d) Runtime vs. quantile summary size.

Fig. 6.4 Efficiency.

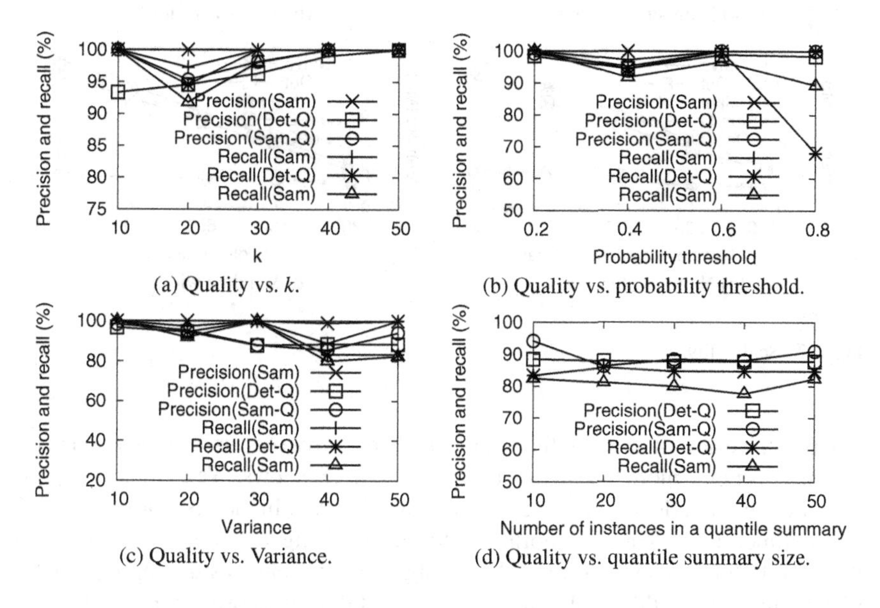

(a) Quality vs. k. (b) Quality vs. probability threshold.

(c) Quality vs. Variance. (d) Quality vs. quantile summary size.

Fig. 6.5 Approximation quality.

6.4.3 Efficiency and Approximation Quality

Figure 6.4 shows the runtime of the four algorithms. To show the effectiveness of the compatible dominant set technique and the pruning techniques discussed in Sec-

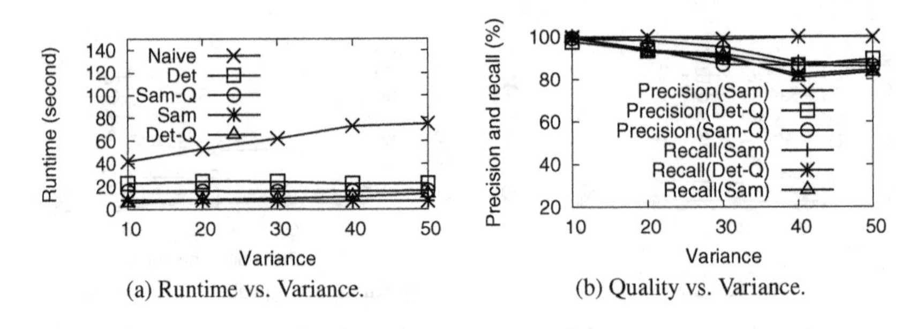

(a) Runtime vs. Variance.

(b) Quality vs. Variance.

Fig. 6.6 Efficiency and approximation quality on data sets under the Gamma distribution.

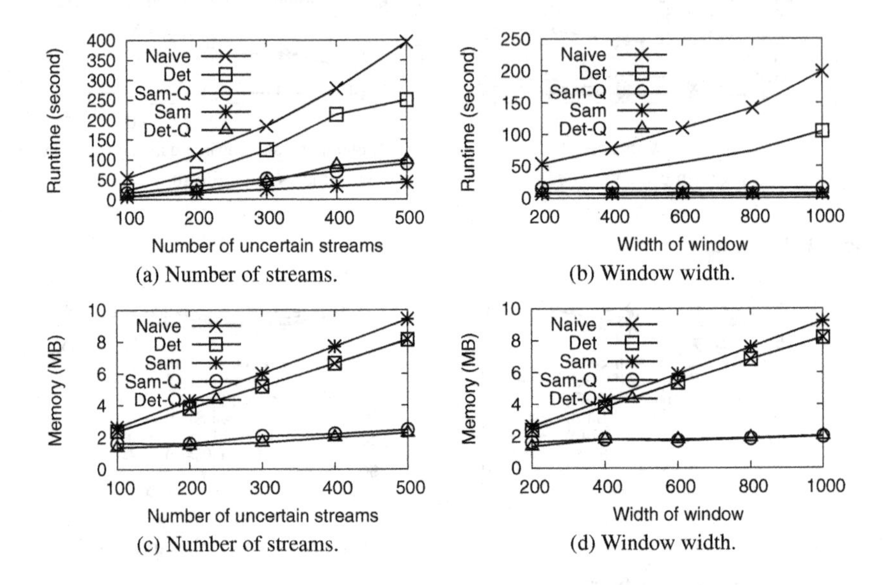

(a) Number of streams.

(b) Window width.

(c) Number of streams.

(d) Window width.

Fig. 6.7 Scalability.

tion 6.1.2, we also plot the runtime of the method (*Naive*) discussed in Section 6.1.1, which does not explore the sharing between sliding windows. We evaluate the probabilistic top-k queries in 5 consecutive sliding windows. In the first sliding window, the runtime of the "Naïve" algorithm and the "Det" algorithm is the same. But in the next 4 sliding windows, the "Naive" algorithm recomputes the top-k probabilities without using the results in the previous sliding windows. But the "Det" algorithm adopts the incremental window maintenance techniques, and thus requires very little computation. Therefore, by average, the runtime of the "Naive" algorithm is approximately 5 times greater than the runtime of the "Det" algorithm. Among all methods, the sampling method and the algorithm using quantiles have much less runtime than the deterministic methods.

Figure 6.4(a) shows that when parameter k increases, the runtime of the naive method and the deterministic method also increases. With a larger k, more instances are likely to be ranked top-k, and thus more instances have to be read before pruning techniques take effects. Moreover, the Poisson binomial recurrence used by those two methods has a linear complexity with respect to k. However, the deterministic method has a clear advantage over the naive method, which shows the effectiveness of the compatible dominant set technique, and the pruning using highest possible rank. The runtime in the sampling methods and the *Det-Q* method is more sensitive to k, since those techniques have very small overhead increase as k increases.

In Figure 6.4(b), as the probability threshold increases, the runtime of the naive method and the deterministic method first increase, and then drop when p is greater than 0.8. As indicated by Theorem 6.1, if p is small, we can determine that many streams can pass the threshold after checking only a small number of their instances; if p is very large, we can also determine that many streams fail the threshold after checking a small number of instances.

In the synthetic data set, the instances of each stream follow a normal distribution. If the variance of the distribution is larger, then the ranks of instances are more diverse. Thus, we may have to scan more instances in order to determine whether the top-k probability of a stream can pass the probability threshold. Figure 6.4(c) verifies our analysis.

We test how the two parameters ϕ and ε affect the efficiency of the methods using quantiles. $\lceil \frac{1}{\phi} \rceil$ is the number of instances kept in a quantile summary. In Figure 6.4(d), we set the value of ϕ to $0.1, 0.05, 0.033, 0.025, 0.02$, and the corresponding number of instances in a quantile summary is $10, 20, 30, 40, 50$, respectively. Only the runtime of *Det-Q* increases when more instances are kept. Since ε is typically very small and does not affect the runtime remarkably, we omit the details here.

Figure 6.5 compares the precision and the recall of the three approximation algorithms using the same settings as in Figure 6.4. In general, all three methods have good approximation quality, and the sampling method achieves a higher precision and recall. We notice that, in Figure 6.5(b), the recall of the deterministic methods using quantiles decreases when the probability threshold increases. This is because a larger probability threshold reduces the size of answer sets. Moreover, in Figure 6.5(c), the precision and recall of all methods decreases slightly as the variance increases. When the variance gets larger, the instances of a stream distribute more sparsely. Thus, we may need more samples to capture the distribution.

We also test our methods on the uncertain data streams generated using the Gamma distribution $\Gamma(k, \theta)$. The mean μ is randomly picked from a domain $[0, 1000]$. We change the variance σ from 10 to 50. The scale parameter θ is set to $\frac{\sigma}{\mu}$ and the shape parameter k is set to $\frac{\mu^2}{\sigma}$. The efficiency and the approximation quality are shown in Figure 6.6. The results are very similar to Figure 6.4(c) and Figure 6.5(c). This shows that the performance of our algorithms is not sensitive to the types of score distributions of the data sets.

6.4.4 Scalability

To test the scalability of the algorithms, we first fix the sliding window width to 200, and change the number of uncertain streams from 100 to 500. The maximum number of instances in a window is $100,000$. As the number of streams increases, the Poisson binomial recurrence takes more time in the deterministic method. Thus, the runtime increases. However, all methods are linearly scalable. The results are shown in Figure 6.7(a).

Then, we fix the number of streams to 100, and vary the sliding window width from 200 to $1,000$. The runtime in the deterministic method increases substantially faster than the other methods. The sampling method is more stable, because its runtime is related to only the sample size and the number of streams. For the methods using quantile summaries, after compressing the instances in to a quantile summary, the increase of sliding window width does not affect the runtime noticeably. The results are shown in Figure 6.7(b).

In terms of memory usage, Figures 6.7(c) and 6.7(d) show the scalability of each algorithm with respect to the number of uncertain streams and the sliding window width, respectively. The memory used by the deterministic exact algorithm and the sampling algorithm increases linearly, since it is proportional to the number of instances in a sliding window. The memory used by the extended deterministic method using quantiles and the sampling method using quantiles does not change dramatically, because the number of instances for each object in the sliding window only depends on parameter ϕ.

6.5 Summary

In this chapter, we proposed a novel uncertain data stream model and continuous probabilistic threshold top-k queries on uncertain streams, which are different from the existing probabilistic stream models and queries. A deterministic method and a sampling method, as well as their space efficient versions using quantiles were developed to answer those queries. Experimental results on real data sets and synthetic data sets were reported, which show the effectiveness and efficiency of the methods.

Chapter 7
Ranking Queries on Probabilistic Linkages

In Chapters 4, 5 and 6, we adopt an independent uncertain object model. That is, we assume that any two uncertain objects in a data set are independent. In many applications, various types of dependencies may exist among real world uncertain objects, such as the case shown in Example 2.12. In this chapter, we study how to answer probabilistic ranking queries on the probabilistic linkage model.

7.1 Review: the Probabilistic Linkage Model

A probabilistic linkage model contains two sets of tuples A and B and a set of linkages $\mathscr{L}$. Each linkage l in $\mathscr{L}$ matches one tuple in A and one tuple in B. For a linkage $l = (t_A, t_B)$, we say l is associated with t_A and t_B. We write $l \in t_A$ and $l \in t_B$.

We can consider each tuple $t_A \in A$ as an uncertain object. An tuple $t_B \in B$ can be considered as an instance of t_A if there is a linkage $l = (t_A, t_B) \in \mathscr{L}$. The membership probability of instance t_B with respect to object t_A is $Pr(l)$. Object t_A may contain multiple instances $\{t_{B_1}, \cdots, t_{B_k}\}$ where $(t_A, t_{B_i}) \in \mathscr{L}$ $(1 \leq i \leq k)$. At the same time, an instance t_B may belong to multiple objects $\{t_{A_1}, \cdots, t_{A_d}\}$ where $(t_{A_j}, t_B) \in \mathscr{L}$ $(1 \leq j \leq d)$. A mutual exclusion rule $R_{t_B} = (t_{A_1}, t_B) \oplus \cdots \oplus (t_{A_d}, t_B)$ specifies that t_B can only belong to one object in a possible world.

Since different objects may share the same instance in the probabilistic linkage model, we develop a probabilistic mutual exclusion graph (PME-graph for short) to describe such dependencies. Moreover, the evaluation methods for probabilistic ranking queries developed for independent uncertain objects cannot be directly applied on the probabilistic linkage model. We develop efficient evaluation algorithms in this chapter.

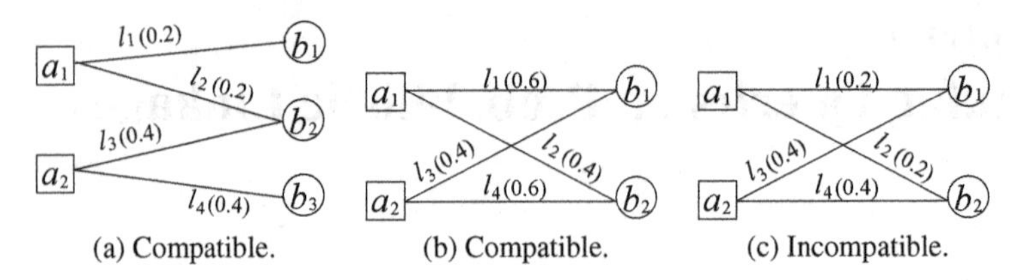

Fig. 7.1 Linkage compatibility.

7.2 Linkage Compatibility

In this section, we study the compatibility of a set of linkages and the effect on the possible world probabilities.

7.2.1 Dependencies among Linkages

The linkage functions defined in Section 2.3.2 give only the probabilities of individual linkages. This is the situation in all state-of-the-art probabilistic linkage methods. In other words, existing linkage methods do not estimate the joint probabilities of multiple linkages. Linkages are not independent – at most one linkage can appear in a possible world among those associated with the same tuple. Then, what roles do dependencies play in defining probabilities of possible worlds?

Example 7.1 (Compatible linkages). Consider the linkages shown in Figure 7.1(a) between tuples in tables $A = \{a_1, a_2\}$ and $B = \{b_1, b_2, b_3\}$. The probabilities of the linkages are labeled in the figure. For a linkage l, let l and $\neg l$ denote the events that l appears and l is absent, respectively. Since linkages l_1 and l_2 are mutually exclusive, the marginal distribution of (l_1, l_2), denoted by $f(l_1, l_2)$, is $Pr(\neg l_1, \neg l_2) = 1 - Pr(l_1) - Pr(l_2) = 0.6$, $Pr(\neg l_1, l_2) = Pr(l_2) = 0.2$, $Pr(l_1, \neg l_2) = Pr(l_1) = 0.2$, and $Pr(l_1, l_2) = 0$. Similarly, the marginal distributions $f(l_2, l_3)$ and $f(l_3, l_4)$ can be calculated from the linkage probabilities and the mutual exclusion rules.

Does there exist a set of possible worlds (i.e., the joint distribution $f(l_1, l_2, l_3, l_4)$) that satisfy the marginal distributions $f(l_1, l_2)$, $f(l_2, l_3)$ and $f(l_3, l_4)$? If so, can we further determine the existence probability of each possible world? The answer is yes in this example. Based on Bayes' theorem, we can compute the joint distribution

$$f(l_1, l_2, l_3, l_4) = f(l_1, l_2) f(l_3 | l_2) f(l_4 | l_3) = \frac{f(l_1, l_2) f(l_2, l_3) f(l_3, l_4)}{f(l_2) f(l_3)}.$$

As another example of compatible linkages, consider Figure 7.1(b). The joint probabilities are $Pr(l_1, \neg l_2, \neg l_3, l_4) = 0.6$ and $Pr(\neg l_1, l_2, l_3, \neg l_4) = 0.4$.

Figure 7.1(c) gives an example of incompatible linkages. Linkages in Figure 7.1(c) have the same mutual exclusion rules as the ones in Figure 7.1(b), but the proba-

bilities are different. From the probability of each linkage and the mutual exclusion rules, we can compute the marginal distributions $f(l_1,l_2)$, $f(l_3,l_4)$, $f(l_1,l_3)$ and $f(l_2,l_4)$, respectively. The three marginal distributions $f(l_1,l_2)$, $f(l_3,l_4)$ and $f(l_2,l_4)$ can uniquely determine a joint distribution $f(l_1,l_2,l_3,l_4)$. Due to the mutual exclusion rule $l_1 \oplus l_3$, the probability that l_1 and l_3 both appear should be 0. However, from this joint distribution, we can derive

$$Pr(l_1,l_3) = Pr(l_1,\neg l_2,l_3,\neg l_4) = Pr(l_1,\neg l_2)Pr(\neg l_4|\neg l_2)Pr(l_3|\neg l_4) = \frac{1}{15}.$$

Thus, the joint probability $f(l_1,l_2,l_3,l_4)$ computed from the marginal distributions $f(l_1,l_2)$, $f(l_3,l_4)$ and $f(l_1,l_3)$ is inconsistent with the marginal distribution $f(l_1,l_3)$. Therefore, the linkages in Figure 7.1(c) are not compatible. ■

Definition 7.1 (Compatible linkages). A set of linkages are *compatible* if there is at least a joint distribution on the linkages that satisfies the marginal distributions specified by the linkages. ■

Example 7.1 indicates that some linkages may lead to a situation where the possible worlds cannot be decided (i.e., the linkages are not compatible). In the rest of this section, we will discuss three problems.

1. In what situations are the linkages compatible?
2. How to fix incompatible linkages to compatible ones with small information loss?
3. How to compute the probabilities of possible worlds for compatible linkages?

7.2.2 Probabilistic Mutual Exclusion Graphs

Dependencies among linkages are important in deriving possible world probabilities. We develop a probabilistic graphic model to capture such dependencies.

Definition 7.2 (PME-graph). Given a set of probabilistic linkages $\mathcal{L}_{A,B}$, a **probabilistic mutual exclusion graph (PME-graph)** $G_{\mathcal{L},A,B} = (V,E)$ is an undirected graph such that (1) a vertex $v_l \in V$ $(l \in \mathcal{L}_{A,B})$ is a binary random variable corresponding to a probabilistic linkage, $Pr(v_l = 1) = Pr(l)$ and $Pr(v_l = 0) = 1 - Pr(l)$; (2) an edge $e = (v_l, v_{l'}) \in E$ $(v_l, v_{l'} \in V)$ if linkages l and l' share a common tuple, i.e., they are involved in a mutual exclusion rule R_t $(t \in A$ or $t \in B)$. ■

A PME-graph may contain multiple connected components. The vertices in one connected component are correlated. The vertices in a PME-graph $G_{\mathcal{L},A,B} = (V,E)$ have several basic properties.

Corollary 7.1 (Vertices in a PME-graph). *For a PME-graph $G = (V,E)$, two vertices $v_i, v_j \in V$ has the following properties:*

1. $v_i, v_j \in V$ are independent if v_i and v_j belong to different connected components.

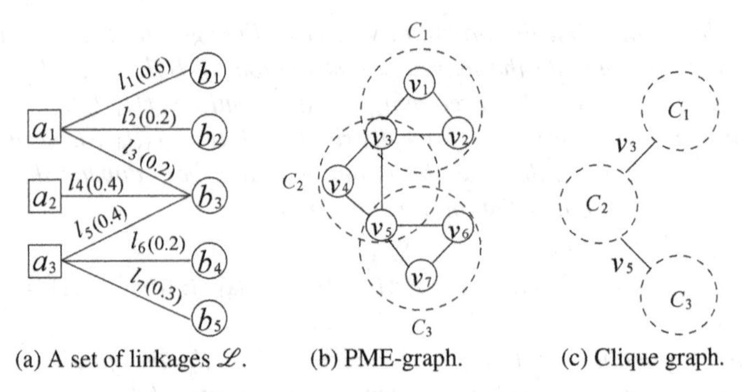

(a) A set of linkages $\mathscr{L}$. (b) PME-graph. (c) Clique graph.

Fig. 7.2 A set of linkages and the corresponding PME-graph and clique graph .

2. *v_i and v_j are mutually exclusive if there is an edge $e = (v_i, v_j) \in E$.*
3. *v_i and v_j are conditionally independent given another vertex v if there is a path between v_i and v_j passing v.*

■

Theorem 7.1. *A PME-graph $G_{\mathscr{L},A,B}$ is a Markov random field.*
Proof. We need to show that $G_{\mathscr{L},A,B}$ satisfies the Markov property, which states that, in a set of random variables, the probability of any given random variable X being in a state x only depends on a subset of random variables in the system [134]. In a PME-graph $G_{\mathscr{L},A,B}$, a vertex v is mutually exclusive with its adjacent vertices N_v. For any other vertex $v' \in V - \{v\} - N_v$, v and v' are independent conditional on N_v. The Markov property is satisfied. ■

A PME-graph has two interesting and useful properties.

Lemma 7.1 (Property of PME-graphs). *For a PME-graph G corresponding to linkages $\mathscr{L}$ between tuple sets A and B:*

1. *For a tuple $t \in A$ or $t \in B$, the edges corresponding to the linkages in the mutual exclusion rule R_t form a maximal clique in G.*
2. *Any two cliques in G can share at most one common vertex.*

Proof. Since a maximal clique in a PME-graph G captures a mutual exclusion rule in the linkages $\mathscr{L}_{A,B}$, the first item holds.

To prove the second item, since a vertex v in $G_{\mathscr{L},A,B}$ is a linkage and can involve only one tuple in A and another tuple in B, v can participate in at most 2 maximal cliques in the PME-graph. ■

For example, two maximal cliques $C_1 = \{v_1, v_2, v_3\}$ and $C_2 = \{v_3, v_4, v_5\}$ in Figure 7.2(b) only share one common vertex v_3.

A PME-graph captures two types of dependencies among linkages: first, two linkages associated with the same tuple are mutually exclusive; second, two linkages are conditionally independent given the linkages connecting the two linkages.

PME-graphs are useful in deriving possible worlds of linkages and ranking query evaluation methods, which will be discussed in Sections 7.2.5, 7.3 and 7.4.

Besides PME-graphs, we create a maximal clique graph to represent the dependencies between maximal cliques. Hereafter, only maximal cliques in the PME-graphs are of interest. For the sake of simplicity, we refer to maximal cliques as cliques.

Definition 7.3 (Clique graph). Given a PME-graph $G_{\mathscr{L},A,B}$, the corresponding **clique graph** is a graph $G_{clique}(V,E)$, where a vertex $v_C \in V$ corresponds to a maximal clique C in $G_{\mathscr{L},A,B}$ and an edge $e_{CC'} = (v_C, v_{C'}) \in E$ if cliques C and C' in the PME-graph share a common vertex.

Let C be a maximal clique in $G_{\mathscr{L},A,B}$ and V_C be the set of vertices in C. The probability of the corresponding vertex $v_C \in V$ in the clique graph is $Pr(v_C) = \sum_{x \in V_C} Pr(x)$. ∎

Hereafter, in order to distinguish between the vertices in a PME-graph and the vertices in a clique graph, we refer to a vertex in a clique graph as a clique.

Figures 7.2(a), (b) and (c) show a set of linkages, the PME-graph and the clique graph, respectively. Each node v_i in Figure 7.2(b) corresponds to a linkage l_i in Figure 7.2(a). Each maximal clique in Figure 7.2(b) corresponds to a vertex in Figure 7.2(c).

7.2.3 Compatibility of Linkages

To check wether a set of linkages $\mathscr{L}$ are compatible, a straightforward approach is to check, for each clique $C \in G_{clique}$, whether the joint probability on $G_{clique} - C$ can lead to a marginal distribution on C that is consistent with the given marginal distribution $f(C)$. However, this approach is very costly since we have to compute the joint probability on cliques $G_{clique} - C$ for every C.

Fortunately, we can derive a sufficient and necessary condition for compatible linkages as stated in the following theorem.

Theorem 7.2 (Compatibility). *Given a set of linkages $\mathscr{L}$ and the corresponding clique graph G_C, then linkages in $\mathscr{L}$ are compatible if and only if, for each connected component $G' \in G_C$, one of the following two conditions holds:*

1. *G' is acyclic;*
2. *G' is a cycle such that each vertex v_C in the cycle is connected to two edges e_1 and e_2, whose corresponding vertices v_1 and v_2 in the PME-graph satisfy $Pr(v_1) + Pr(v_2) = 1$.*

Proof. We first prove the sufficiency, that is, if the clique graph of a set of linkages satisfies one of the two conditions in the theorem, then the linkages are compatible.

Condition 1. If the clique graph G_C is acyclic, then the joint distribution of the linkages can be derived using the methods discussed in Section 7.2.5.

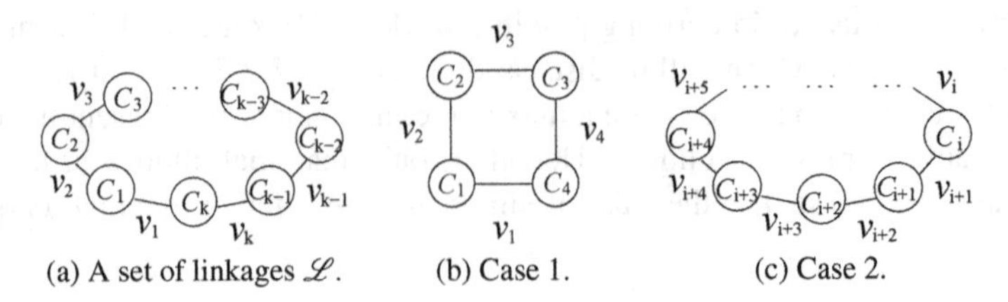

(a) A set of linkages $\mathscr{L}$. (b) Case 1. (c) Case 2.

Fig. 7.3 A cycle of k cliques.

Condition 2. If the second condition holds, then the joint distribution of the linkages involved in G' can be uniquely determined. Suppose G' contains vertices $\{v_{C_1},\cdots,v_{C_k}\}$ as shown in Figure 7.3(a), whose corresponding cliques are $\{C_1,\cdots,C_k\}$ in the PME-graph (k must be an even number since the linkages between tuple sets A and B form a bipartite graph). In the clique graph, since each vertex v_{C_i} has degree 2, which means, the corresponding clique in the PME-graph shares 2 vertices with 2 other cliques. There are k edges $e_{v_1},\cdots,e_{v_k}$ involved in the cycle, whose corresponding vertices in the PME-graph are $v_1,\cdots,v_k$. Each vertex v_i belongs to two cliques C_{i-1} and C_i ($2\leq i\leq k$). v_1 belongs to cliques C_1 and C_k. Since the probability sum of each two connected edges is 1, we have $Pr(v_1)=Pr(v_{2i+1})$ ($1\leq i\leq \frac{k}{2}-1$) and $1-Pr(v_1)=Pr(v_{2j})$ ($1\leq j\leq \frac{k}{2}$). Thus, the joint distribution of all vertices in the PME-graph is given by

$$Pr((\bigwedge_{0\leq i\leq \frac{k}{2}-1} v_{2i+1})\bigwedge(\bigwedge_{1\leq j\leq \frac{k}{2}} \neg v_{2j}))=Pr(v_1)$$
$$Pr((\bigwedge_{0\leq i\leq \frac{k}{2}-1} \neg v_{2i+1})\bigwedge(\bigwedge_{1\leq j\leq \frac{k}{2}} v_{2j}))=1-Pr(v_1).$$

The joint distribution is consistent with the marginal distribution specified by each linkage. Thus, the linkages are compatible.

Then, we prove the necessity. That is, if a set of linkages are compatible, then the corresponding clique graph must satisfy one of the two conditions in the theorem. Consider a set of compatible linkages whose clique graph is G_C.

Suppose G' contains a cycle, then we need to show that G' can only form a cycle satisfying condition 2 in the theorem. We prove this in two cases: the cycle contains 4 vertices and the cycle contains more than 4 cases.

Case 1: The cycle in G' contains 4 vertices $v_{C_1},v_{C_2},v_{C_3},v_{C_4}$, whose corresponding cliques in the PME-graph are $\{C_1,C_2,C_3,C_4\}$, as illustrated in Figure 7.3(b). Let v_i be the vertex contained by C_{i-1} and C_i ($2\leq i\leq 4$) and v_1 be contained by C_1 and C_4. The joint probability of v_1 and v_4 can be expressed as

$$Pr(v_1v_4)=Pr(\neg v_2\neg v_3)Pr(v_1|\neg v_2)Pr(v_4|\neg v_3).$$

Since v_1 and v_4 are contained in clique C_4, $Pr(v_1v_4)=0$ holds. Moreover, $Pr(v_1|\neg v_2)>0$ and $Pr(v_4|\neg v_3)>0$. Thus, $Pr(\neg v_2\neg v_3)=0$. Since v_2 and v_3 are contained in the same clique C_2, we have $Pr(\neg v_2\neg v_3)=1-Pr(v_2)-Pr(v_3)$. Therefore, $Pr(v_2)+Pr(v_3)=1$, which means that C_2 only contains two vertices $\{v_2,v_3\}$

and $Pr(C_2) = 1$. Similarly, we can show that other clique C_i ($1 \leq i \leq 4$) only contains 2 vertices and the probability sum of the two vertices is 1.

Case 2: The cycle in G' contains k vertices $v_{C_1}, \cdots, v_{C_k}$ ($k > 4$), as illustrated in Figure 7.3(c). The corresponding cliques in the PME-graph are $C_1, \cdots, C_k$, respectively. Let v_i be the vertex contained by C_{i-1} and C_i ($2 \leq i \leq k$) and v_1 be contained by C_1 and C_k. We show that, for any clique C_i, $Pr(v_i) + Pr(v_{i+1}) = 1$.

The joint distribution of v_{i+2} and v_{i+3} can be expressed as

$$Pr(v_{i+2}v_{i+3}) = Pr(\neg v_{i+1}\neg v_{i+4})Pr(v_{i+2}|\neg v_{i+1})Pr(v_{i+3}|\neg v_{i+4})$$

Since v_{i+2} and v_{i+3} belong to the same clique C_{i+2}, we have $Pr(v_{i+2}v_{i+3}) = 0$. Moreover, $Pr(v_{i+2}|\neg v_{i+1}) > 0$ and $Pr(v_{i+3}|\neg v_{i+4}) > 0$. Therefore, $Pr(\neg v_{i+1}\neg v_{i+4}) = 0$. We can express $Pr(\neg v_{i+1}\neg v_{i+4})$ as

$$\begin{aligned} Pr(\neg v_{i+1}\neg v_{i+4}) = {} & Pr(\neg v_{i+1}\neg v_{i+4}v_iv_{i+5}) + Pr(\neg v_{i+1}\neg v_{i+4}v_i\neg v_{i+5}) \\ & + Pr(\neg v_{i+1}\neg v_{i+4}\neg v_iv_{i+5}) + Pr(\neg v_{i+1}\neg v_{i+4}\neg v_i\neg v_{i+5}) = 0 \end{aligned} \quad (7.1)$$

Since all probability values are non-negative, each component in Equation 7.1 has to be 0. Therefore, we have

$$Pr(\neg v_{i+1}\neg v_{i+4}v_iv_{i+5}) = Pr(v_iv_{i+5})Pr(\neg v_{i+1}|v_i)Pr(\neg v_{i+4}|v_{i+5}) = 0 \quad (7.2)$$

and

$$Pr(\neg v_{i+1}\neg v_{i+4}\neg v_iv_{i+5}) = Pr(\neg v_iv_{i+5})Pr(\neg v_{i+1}|\neg v_i)Pr(\neg v_{i+4}|v_{i+5}) = 0 \quad (7.3)$$

In Equation 7.2, since $Pr(\neg v_{i+1}|v_i) = Pr(\neg v_{i+4}|v_{i+5}) = 1$, we have $Pr(v_iv_{i+5}) = 0$. Therefore, in Equation 7.3, $Pr(\neg v_iv_{i+5}) = Pr(v_{i+5}) - Pr(v_iv_{i+5}) > 0$. Thus, $Pr(\neg v_{i+1}|\neg v_i) = 0$. Since $Pr(\neg v_{i+1}|\neg v_i) = \frac{1-Pr(v_{i+1})-Pr(v_i)}{Pr(\neg v_i)}$, we have $Pr(v_{i+1}) + Pr(v_i) = 1$. Therefore, C_i only has 2 vertices $\{v_i, v_{i+1}\}$ and $Pr(v_{i+1}) + Pr(v_i) = 1$. ∎

For example, Figures 7.1(a) and (b) satisfy the first and the second conditions in Theorem 7.2, respectively. The links there are compatible. Figure 7.1(c) does not satisfy Theorem 7.2, thus, the linkages there are not compatible.

7.2.4 Resolving Incompatibility

Given a set of incompatible linkages $\mathscr{L}$, intuitively, we want to find a subset of $\mathscr{L}$ such that the loss of information is minimized.

How can we measure the amount of information that is retained in a subset of linkages? Different definitions can be adopted in different applications. For instance, in Example 2.12, if medical experts are more interested in the patients in the hospitalization registers, then we may want to select a subset of compatible linkages such that the expected number of patients who are linked to another record in the causes-of-death registers is maximized.

If we do not have any specific application requirement, then it is intuitive to maximize the expected number of linkages in the subset, since it is the information available for analysis. The problem of finding the maximized compatible linkage set is defined as follows.

Definition 7.4 (Maximum compatible linkage set). Given a set of incompatible linkages $\mathscr{L}$, the maximum compatible linkage set is a subset of linkages $\mathscr{L}' \subset \mathscr{L}$ such that:

1. All linkages in $\mathscr{L}'$ are compatible, and
2. The expected number of linkages in $\mathscr{L}'$ is maximum over all compatible linkage subsets of $\mathscr{L}$. That is, we define a binary random variable X_l for each linkage $l \in \mathscr{L}$ such that $Pr(X_l = 1) = Pr(l)$ and $Pr(X_l = 0) = 1 - Pr(l)$. $\mathscr{L}' = \arg\max_{L \subset \mathscr{L}} E[\sum_{l \in L} X_l]$, where $E[\sum_{l \in L} X_l]$ is the expected number of linkages in L.

■

In order to find the maximum compatible linkage set in $\mathscr{L}$, a naïve approach is to enumerate all compatible linkage subsets of $\mathscr{L}$ and compute the expected number of linkages in each subset. The subset with the maximum expected number of linkages is returned. We conject that this problem is NP-hard.

We can apply an approximation approach as follows. Let G_C be the clique graph of $\mathscr{L}$, each vertex v_C in G_C is corresponding to a clique C in the PME-graph G. Each edge e_v in between two vertices v_{C_1} and v_{C_2} G_C is corresponding to the common vertex v of the two cliques C_1 and C_2 in G. Let the weight of edge e_v be the probability of the corresponding vertex v. Then, we can find the *maximum spanning forest* [190] of the graph G_C. A maximum spanning tree of a connected component in G_C is a spanning tree that has the maximum sum of edge weights. Intuitively, the maximum spanning tree excludes the edges in the clique graph with small weights (that is, the probability values of the corresponding vertices). Therefore, the corresponding linkages may have a large expected number of linkages.

Since the main focus of our study is ranking queries on probabilistic linkages, we assume that the linkages are compatible hereafter.

7.2.5 Deriving All Possible Worlds

To enumerate all possible worlds of a set of linkages $\mathscr{L}_{A,B}$, a naïve approach is to check each subset of the linkages against Definition 2.20. The naïve method takes $O(2^{|\mathscr{L}_{A,B}|})$ time. However, the actual number of valid possible worlds may be much smaller than $2^{|\mathscr{L}_{A,B}|}$. For example, in Figure 7.2(a), there are 7 linkages but there are only 11 possible worlds.

In this section, we use the PME-graph to generate all possible worlds in $O(|\mathscr{W}|)$ time, where $|\mathscr{W}|$ is the number of possible worlds.

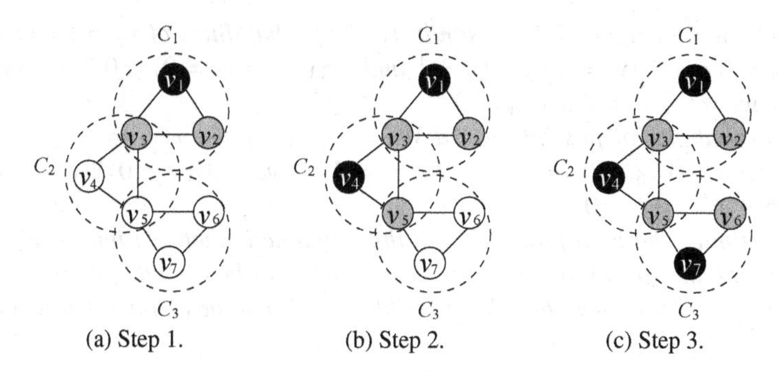

Fig. 7.4 Computing the probability for possible world $W = \{v_1, v_4, v_7\}$. Black and grey vertices are assigned values 1 and 0, respectively. White vertices have not been assigned any value yet.

A possible world W of linkages $\mathscr{L}_{A,B}$ can be regarded as an assignment of values 0 and 1 to the vertices in the PME-graph $G_{\mathscr{L},A,B}$, where a vertex $v_l = 1$ if the corresponding linkage $l \in W$, otherwise $v_l = 0$. For a clique C in $G_{\mathscr{L},A,B}$, if $\sum_{v \in C} Pr(v = 1) < 1$, then at most one vertex in C can be assigned to 1; if $\sum_{v \in C} Pr(v = 1) = 1$, then there is exactly one vertex in C taking value 1. The probability of a possible world W is the joint distribution

$$Pr(W) = Pr((\wedge_{l \in W} v_l = 1) \wedge (\wedge_{l' \notin W} v_{l'} = 0)). \tag{7.4}$$

Since vertices in different connected components in $G_{\mathscr{L},A,B}$ are independent (Section 7.2.2), if $G_{\mathscr{L},A,B} = (V, E)$ contains k connected components $V = V_1 \cup V_2 \cup \cdots V_k$, Equation 7.4 can be rewritten as

$$Pr(W) = \prod_{i=1}^{k} \Pr((\bigwedge_{l \in W \cap V_i} v_l = 1) \bigwedge (\bigwedge_{l' \notin W, l' \in V_i} v_{l'} = 0)) \tag{7.5}$$

The remaining question is how to generate the possible worlds in one connected component.

Example 7.2 (Factoring joint probability). Consider the set of linkages in Figure 7.2(a) and the corresponding PME-graph in Figure 7.2(b). Figure 7.4 shows how we generate a possible world using the PME-graph.

We consider cliques C_1, C_2, C_3 in the PME-graph one by one. In the first step, since no value has been assigned in clique C_1, all three vertices in C_1 may have a chance to be set to 1. The probability of v_1 taking value 1 is $Pr(v_1 = 1) = 0.6$. Suppose we set $v_1 = 1$. Then, v_2 and v_3 can only take value 0 (Figure 7.4(a)).

After clique C_1 is set, we consider clique C_2. Since $v_3 = 0$, only v_4 and v_5 can take value 1. Moreover, given condition $v_3 = 0$, the probabilities of $v_4 = 1$ and $v_5 = 1$, respectively, are $Pr(v_4 = 1 | v_3 = 0) = \frac{Pr(v_4=1)}{Pr(v_3=0)} = 0.5$ and $Pr(v_5 = 1 | v_3 = 0) = 0.5$. Suppose we set $v_4 = 1$. Then, $v_5 = 0$ (Figure 7.4(b)).

For clique C_3 (Figure 7.4(c)), similarly, the probabilities of $v_6 = 1$ and $v_7 = 1$ when $v_5 = 0$ are $Pr(v_6 = 1|v_5 = 0) = \frac{1}{3}$ and $Pr(v_7 = 1|v_5 = 0) = 0.5$, respectively. Suppose we set $v_8 = 1$. Then, $v_6 = 0$.

The probability of possible world $W_1 = \{l_1, l_4, l_7\}$ is $Pr(v_1 = 1, v_2 = 0, v_3 = 0, v_4 = 1, v_5 = 0, v_6 = 0, v_7 = 1) = Pr(v_1 = 1)Pr(v_4 = 1|v_3 = 0)Pr(v_7 = 1|v_5 = 0) = 0.6 \times 0.5 \times 0.5 = 0.15$.

Interestingly, the joint probability in this example is factored into a set of conditional probabilities on a subset of vertices that can be directly derived from the given linkages. Moreover, the other possible worlds can be enumerated recursively in the same way. ■

7.3 Ranking Queries on Probabilistic Linkages

We formulate the probabilistic ranking queries introduced in Section 2.2.2 on probabilistic linkages as follows.

Given a set of linkages $\mathscr{L}_{A,B}$ between tables A and B, let $Q^k_{P,f}$ be a top-k selection query, where P is a predicate and f is a scoring function which may involve attributes in A, B, or both. For a linkage $l \in \mathscr{L}_{A,B}$, $f(l)$ is a real value score.

Definition 7.5 (Top-k probability of linkages and tuples). For a linkage $l \in \mathscr{L}_{A,B}$, the **rank-k probability** of l, denoted by $Pr(l,k)$, is the probability that l is ranked at the k-th position in possible worlds. That is

$$Pr(l,k) = \sum_{W \in \mathscr{W}_{\mathscr{L}_{A,B}}, W_f(k) = l} Pr(W)$$

where $W_f(k)$ is the k-th linkage in possible world W. Moreover, the **top-k probability** of l is

$$Pr^k(l) = \sum_{i=1}^{k} Pr(l,i)$$

Consequently, the **rank-k probability** of a tuple $t \in A \cup B$ is

$$Pr(t,k) = \sum_{l \in t} Pr(l,k)$$

The **top-k probability** of t is

$$Pr^k(t) = \sum_{l \in t} Pr^k(l)$$

■

Given a positive integer k and probability threshold $p \in (0,1]$, a **probabilistic threshold top-k query (Definition 2.12) on linkages** finds the probabilistic linkages whose top-k probabilities are at least p. Similarly, a **probabilistic threshold top-k query on tuples** finds the tuples whose top-k probabilities are at least p.

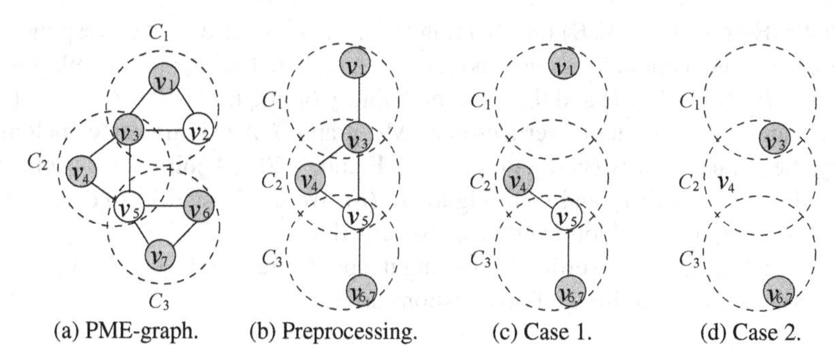

(a) PME-graph. (b) Preprocessing. (c) Case 1. (d) Case 2.

Fig. 7.5 A chain of cliques.

Due to the dependencies among tuples, the query answering techniques developed in Chapter 5 cannot be directly applied to answering probabilistic threshold top-k queries on the probabilistic linkage model. There are four major differences, as discussed in Sections 7.3.1 to 7.3.4, respectively.

7.3.1 Predicate Processing

To answer a top-k selection query $Q^k_{P,f}$, the first step is to deal with the query predicate P. The predicate processing technique used for probabilistic linkages is different from the technique used for independent uncertain objects.

As discussed in Section 5.1, given an uncertain table T and a top-k selection query $Q^k_{P,f}$, we can apply the query predicate P as preprocessing. That is, we select all tuples satisfying the query predicate as $P(T) = \{t | t \in T \wedge P(t) = true\}$. The problem of answering the PT-k query on T can be transformed into finding the tuples in $P(T)$ whose top-k probability values pass the probability threshold.

The same preprocessing does not work on the probabilistic linkage model due to dependencies, as illustrated in the following example.

Example 7.3 (Processing query predicates). Consider the linkages in Figure 7.2. Suppose a predicate P selects linkages $\{l_1, l_3, l_4, l_6, l_7\}$. l_2 and l_5 do not satisfy the predicate P. We use a PME-graph to represent the linkages and use the shaded nodes to represent the linkages satisfying predicate P, as shown in Figure 7.5(a). If we adopt the same preprocessing method described in Section 5.1, then, we should remove vertices v_2 and v_5 that do not satisfy P.

However, by removing vertex v_5, the vertices in C_3 and those in C_2 become disconnected, which means that they are independent. Then, the dependencies among vertices are not preserved.

At the same time, removing v_2 does not change the dependencies among vertices, and thus, does not affect the answers to the query. ∎

In a PME-graph $G = (V, E)$ representing linkages $\mathscr{L}$, a vertex v is corresponding to a linkage l and a clique C is corresponding to a tuple t. The top-k probability of a vertex v is $Pr^k(v) = Pr^k(l)$ and the top-k probability of clique C is $Pr^k(C) = Pr^k(t)$.

There are two categories of vertices in a PME-graph G. A **private vertex** belongs to only one clique, such as $\{v_1, v_2, v_4, v_6, v_7\}$ in Figure 7.2(b). A **joint vertex** belongs to two cliques, such as v_3 and v_5 in Figure 7.2(b). We use V_P and V_J to denote the set of private vertices and joint vertices, respectively.

Let V_S denote the set of vertices satisfying the predicate P and $\overline{V_S} = V - V_S$ be the set of vertices not satisfying P. Two questions arise.

Question 1: Which vertices in $\overline{V_S}$ can be removed?

We can partition the vertices in $\overline{V_S}$ into two subsets.

- The first subset contains the joint vertices in $\overline{V_S}$ that lie in at least one path between two vertices in V_S. That is,

$$V_1 = \{v | v \in \overline{V_S} \cap V_J \wedge \exists P = \langle v_1, \cdots, v, \cdots, v_2 \rangle \ s.t. \ v_1 \in V_S \wedge v_2 \in V_S\}.$$

  The vertices in V_1 cannot be removed in preprocessing, since removing them may not preserve the dependencies among the vertices satisfying P.
- The second subset is $V_2 = \overline{V_S} - V_1$. Removing the vertices in V_2 does not change the top-k probabilities of the vertices satisfying P.

Hereafter, let us assume that V_2 has been removed from G as preprocessing. Therefore, two sets of vertices remain in G: V_S contains all vertices satisfying predicate P; V_1 contains all vertices not satisfying P but connecting to vertices in V_S.

Question 2: Should we treat the remaining vertices equally?

Consider two vertices $v_1 \in V_1$ and $v_2 \in V_S$. Given another vertex $v \in V_S$, let $v_1 \preceq_f v$ and $v_2 \preceq_f v$. In a possible world W, whether v_1 appears or not does not affect the rank of v. However, whether v_2 appears or not does change the rank of v.

In order to distinguish between the vertices in V_S and V_1, we associate a flag attribute F with each vertex $v \in G$. If $P(v) = true$, then $v.F = 1$, otherwise $v.F = 0$.

7.3.2 Dominant Subgraphs

The dominant set property (Theorem 5.1) indicates that, to compute the top-k probability of a tuple t in a probabilistic table, we only need to consider the tuples ranked higher than t.

However, to compute the top-k probability of a vertex v in a PME-graph G, only considering the vertices ranked higher than v may lead to incorrect results. This is because, a vertex v' ranked lower than v may connect the two vertices v_1 and v_2 that are ranked higher than v, therefore, not considering v' may not preserve the dependency between v_1 and v_2.

We define a dominant graph of v as follows.

Definition 7.6 (Dominant subgraph of a linkage). For a PME-graph $G = (V, E)$ and a vertex $v \in V$, the **dominant subgraph** of v, denoted by $G(v)$, is the minimum subgraph in G that contains the vertices $v' \in V$ satisfying:

1. $v' \preceq_f v$, or
2. $v' \in V_J$ and v' is in a path $P = \langle v_1, \cdots, v', \cdots, v_2 \rangle$ in G, where $v_1 \preceq_f v$ and $v_2 \preceq_f v$.

■

Theorem 7.3 (The dominant graph property). *For a PME-graph $G = (V, E)$ and a vertex $v \in V$, $Pr^k_{Q,G}(v) = Pr^k_{Q,G(v)}(v)$, where $Pr^k_{Q,G}(v)$ and $Pr^k_{Q,G(v)}(v)$ are the top-k probabilities of v computed on G and in the dominant subgraph $G(v)$ of v, respectively.*

Proof. If a vertex v' is not in $G(v)$, then v' does not appear in any path between the two vertices ranked higher than v. Therefore, the joint probability distribution of the vertices ranked higher than v does not depend on v', which means whether v' appears or not does not affect the top-k probability of v. ■

Similar to the discussion in Section 7.3.1, although all vertices in $G(v)$ need to be considered when computing the top-k probability of v, only the vertices in $G(v)$ that are ranked higher than v affect the actual rank of v. Therefore, we change the assignment of the flag attribute value of each vertex in $G(v)$ as follows. For a vertex $v' \in G(v)$, if $P(v') = true$ and $v' \preceq_f v$, we set its flag $v'.F = 1$, otherwise, we set $v'.F = 0$.

Given a vertex v, how can we obtain its dominant subgraph $G(v)$? Since a PME-graph G may contain multiple connected components, we can construct, in each component G_j, the dominant subgraph $G_j(v)$ as follows.

We traverse the clique graph of G_j in the depth first order. For each clique C, we visit its vertices one by one. If a vertex $v' \in C$ satisfies the condition $v' \preceq_f v$, then we include v' into $G_j(v)$. Moreover, we find the path from C to the last visited clique C' whose vertices are included in $G_j(v)$. All vertices joining the cliques along the path are included in $G_j(v)$ too. Since each vertex is at most visited twice during the construction, the complexity of constructing $G_j(v)$ is $O(|V_{G_j}|)$ where $|V_{G_j}|$ is the number of vertices in G_j.

7.3.3 Vertex Compression

For a set of private vertices $V_p = \{v_{c_1}, \cdots, v_{c_m}\}$ in $G(v)$ belonging to the same clique C, if for any $v' \in V_p$, $v' \preceq_f v$, then we can replace them with a single vertex v_p where

$Pr(v_p) = \sum_{1 \leq i \leq m} Pr(v_{c_i})$. Moreover, for all other vertices $v \in V_C - \{v_{c_1}, \cdots, v_{c_m}\}$, an edge (v, v_p) is added to E. This technique is called *vertex compression.*

Vertex compression does not change the top-k probability of v. The reason is that $\{v_{c_1}, \cdots, v_{c_m}\}$ only belongs to C and assigning 1 to different vertices in $\{v_{c_1}, \cdots, v_{c_m}\}$ does not affect the value assignment of other vertices not in C. After vertex compression, each clique in $G(v)$ contains at most one private vertex.

For example, for a dominant graph $G(v)$ shown in Figure 7.5(a), let the grey vertices be the ones ranked higher than v. After the vertex compression, $G(v)$ is changed to the graph in Figure 7.5(b).

7.3.4 Subgraph Probabilities

Given a dominant subgraph $G(v)$, v is ranked at the j-th position if and only if v appears and there are $j-1$ vertices in $G(v)$ appear.

We define *subgraph probability* $Pr(G(v), j)$ as the probability that j vertices in $G(v)$ appear in possible worlds. Since v and the vertices in $G(v)$ may not be independent, the probability that j vertices in $G(v)$ appear may depend on v. Therefore, we further define the *conditional subgraph probability* of $G(v)$ given v.

Definition 7.7 (Conditional subgraph probability). Given a top-k selection query $Q^k_{Q,f}$, a dominant graph $G(v)$ of a vertex v, the **subgraph probability** $Pr(G(v), i)$ is the probability that i vertices satisfying predicate P appear in $G(v)$, that is,

$$Pr(G(v), i) = \sum_{W \in \mathcal{W}, |\{v' | v' \in W \cap G(v), v'.F = 1\}| = i} Pr(W).$$

Moreover, the **conditional subgraph probability** $Pr(G(v), i|v)$ is the probability that i vertices satisfying predicate P appear in $G(v)$ given the condition that v appears, that is,

$$Pr(G, i|v) = \frac{\sum_{W \in \mathcal{W}, |\{v' | v' \in W \cap G, v'.F = 1\}| = i, v \in W} Pr(W)}{\sum_{W \in \mathcal{W}, v \in W} Pr(W)}$$

■

The top-k probability of v can be computed as

$$Pr^k(v) = Pr(v) \cdot \sum_{i=0}^{k-1} Pr(G(v), i|v). \tag{7.6}$$

Given a dominant graph $G(v)$, the Poisson binomial recurrence (Theorem 5.2) cannot be used to compute the subgraph probabilities, since it only works for independent uncertain objects. In Section 7.4, we will discuss how to compute subgraph probabilities of $G(v)$.

7.4 Tree Recurrence: Subgraph Probability Calculation

The dominant subgraph $G(v)$ of a vertex v may contain multiple connected components. The vertices in different components are independent. Therefore, we first focus on computing the subgraph probabilities of each components $G_i(v)$. Then, the overall subgraph probabilities can be computed from the convolution of the subgraph probabilities of all components.

The cliques in each component $G_i(v)$ form a tree (as discussed in Section 7.2.5). We scan the cliques in $G_i(v)$ in the depth first order. During the scan, we compute subgraph probability of the scanned subtree based on an important property below.

Corollary 7.2 (Conditional independency). *Given a PME-graph G and its clique tree T, for any clique C in G_C, C partition G_C into multiple subtrees $T_1, \cdots, T_m$ and any two vertices in different subtrees are conditionally independent given C.* ■

To compute the conditional subgraph probability $Pr(G(v)|v)$, there are two cases, depending on whether v is in $G_i(v)$ or not. We first discuss the case where v is not in $G_i(v)$, then $Pr(G_i(v), j|v) = Pr(G_i(v), j)$. The second case where v is in $G_i(v)$ will be discussed at the end of Section 7.4.2, since it is a straightforward extension of the first case.

7.4.1 A Chain of Cliques

First, let us consider the simple case where all cliques $C_1, \cdots, C_m$ in $G_i(v)$ form a chain.

Example 7.4 (A chain of cliques). Consider $G_i(v)$ in Figure 7.5(b). Let the darkened vertices be the vertices satisfying the query predicate and ranked higher than v. We want to compute the subgraph probabilities for subgraphs C_1, $C_1 \cup C_2$, and $C_1 \cup C_2 \cup C_3$. The subgraph probabilities for C_1 is simply $Pr(C_1, 0) = 1 - Pr(v_1) - Pr(v_3) = 0.2$ and $Pr(C_1, 1) = 0.8$.

We consider computing the probability $Pr(C_1 \cup C_2, 2)$. All vertices in C_1 and C_2 are conditionally independent given v_3. We consider two cases as illustrated in Figures 7.5(c) and (d), respectively.

***Case 1**: v_3 does not appear. The probability for this case is $Pr(\neg v_3) = 0.8$. Under this condition, v_1 and v_4 are independent. The probabilities that v_1 are v_4 appear in this case are $Pr(v_1|\neg v_3) = \frac{0.6}{0.8} = 0.75$ and $Pr(v_4|\neg v_3) = \frac{0.4}{0.8} = 0.5$, respectively. The probability that v_1 and v_4 both appear is $Pr(v_1, v_4|\neg v_3) = Pr(v_1|\neg v_3)Pr(v_4|\neg v_3) = 0.375$. Therefore, $Pr(C_1 \cup C_2, 2|\neg v_3)$, the conditional probability that 2 vertices appear in $C_1 \cup C_2$ when v_3 does not appear, is 0.375.*

***Case 2**: v_3 appears, with probability $Pr(v_3) = 0.2$. Then, the probability that $Pr(C_1 \cup C_2, 2)$ is 0 since no other vertices in C_1 and C_2 can appear.* ■

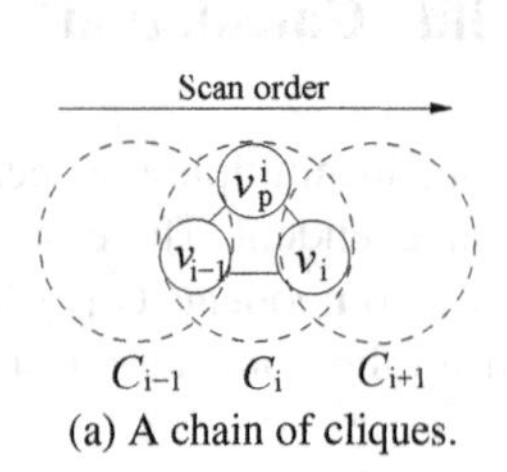

(a) A chain of cliques.

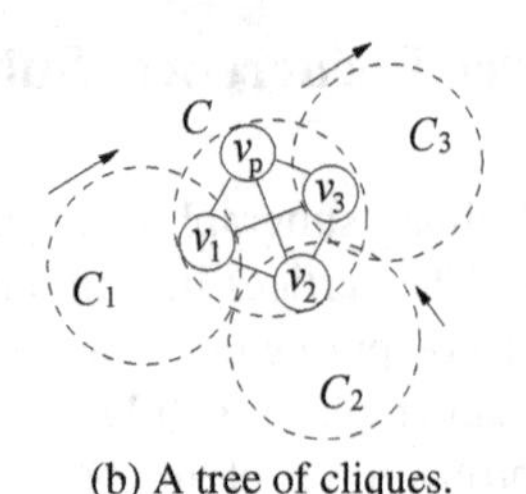

(b) A tree of cliques.

Fig. 7.6 Difference cases in subgraph recurrence.

Generally, let $C_1, \cdots, C_m$ be a chain of cliques. The two consecutive cliques C_{i-1} and C_i share a common vertex v_{i-1} $(2 \leq i \leq m)$. Each clique C_i contains at most three vertices, as illustrated in Figure 7.6(a).

- A private vertex v_p^i only belongs to C_i.
- A head vertex v_{i-1} belongs to C_i and C_{i-1}.
- A tail vertex v_i belongs to C_i and C_{i+1}.

Trivially, C_1 has no head vertex and C_m has no tail vertex.

Let $G_{i-1} = \bigcup_{1 \leq j \leq i-1} C_j$ and $G_i = S_{i-1} \cup C_i$. To compute $Pr(G_i, x | \neg v_i)$, two cases may arise, since $C_i - \{v_{i-1}\}$ and $G_{i-1} - \{v_{i-1}\}$ are conditionally independent given v_{i-1}.

1. v_{i-1} appears. Then no other vertices in C_i can appear.
2. v_{i-1} does not appear. In order to have x selected vertices appear in G_i, we consider two subcases.

   a. no selected vertex in C_i appears. Then there must be x selected vertices in G_{i-1} appear.
   b. 1 vertex in C_i appears. Then there must selected $x-1$ vertices in G_{i-1} appear.

Summarizing the above cases, we have the following theorem.

Theorem 7.4 (Conditional subgraph probabilities). *Given a chain of cliques $C_1, \cdots, C_m$, let $G_i = \bigcup_{1 \leq j \leq i} C_j$, $\{v_{i-1}\} = C_{i-1} \cap C_i$, and v_p^i be the private vertex of $C-i$. The conditional subgraph probability of G_1 given $\neg v_1$ is*

$$Pr(G_1, x | \neg v_1) = \begin{cases} \frac{1 - Pr(v_p^1) - Pr(v_1)}{1 - Pr(v_1)}, & x = 0; \\ \frac{Pr(v_p^1)}{1 - Pr(v_1)}, & x = 1; \\ 0, & x > 1. \end{cases}$$

The conditional subgraph probability of G_1 given v_1 is

$$Pr(G_1, x | v_1) = \begin{cases} 1, \; x = 1; \\ 0, \; x \neq 1. \end{cases}$$

For $1 < i < m$, the conditional subgraph probability of G_i given $\neg v_i$ is

$$\begin{aligned} Pr(G_i,x|\neg v_i) &= Pr(\neg v_{i-1}|\neg v_i)\cdot Pr(G_{i-1},x|\neg v_{i-1})\cdot Pr(\neg v_p^i|\neg v_{i-1}\neg v_i) \\ &+Pr(\neg v_{i-1}|\neg v_i)\cdot Pr(G_{i-1},x-v_p^i.F|\neg v_{i-1})\cdot Pr(v_p^i|\neg v_{i-1}\neg v_i) \\ &+Pr(v_{i-1}|\neg v_i)\cdot Pr(G_{i-1},x|v_{i-1}) \end{aligned}$$

The conditional subgraph probability of G_i given v_i is

$$Pr(G_i,x|v_i) = Pr(G_{i-1},x-v_i.F|\neg v_{i-1})$$

■

To compute subgrph probabilities, we scan cliques $C_1,\cdots,C_m$ and calculate $Pr(G_i,x|v_i)$ and $Pr(G_i,x|\neg v_i)$ for $0 \leq x \leq i$. Then, we compute $Pr(G_m,x)$ for $G_m = \bigcup_{1\leq j\leq m} C_j$ using

$$\begin{aligned} Pr(G_m,x) &= Pr(v_{m-1})Pr(G_{m-1},x|v_{m-1}) \\ &+Pr(\neg v_{m-1})Pr(G_{m-1},x|\neg v_{m-1})Pr(\neg v_p^m|\neg v_{m-1}) \\ &+Pr(\neg v_{m-1})Pr(G_{m-1},x-v_p^m.F|\neg v_{m-1})Pr(v_p^m|\neg v_{m-1}) \end{aligned} \tag{7.7}$$

The overall complexity is $O(m^2)$.

7.4.2 A Tree of Cliques

Now let us consider the general case where the clique graph of $G_i(v)$ is a tree.

Example 7.5 (Connecting multiple cliques). Figure 7.6(b) shows three cliques C_1, C_2 and C_3 connected by clique C. C_1, C_2 and C_3 are the end vertices of three clique chains. Let v_1, v_2 and v_3 be the three common vertices between C and C_1, C_2 and C_3, respectively. Then, there are five possible value assignments of C: $v_p = 1$, $v_1 = 1$, $v_2 = 1$, $v_2 = 1$ or no vertices taking value 1 (if $\sum_{v\in V_C} Pr(v) < 1$). In any of the five cases, the joint probability distribution of C_1, C_2 and C_3 are conditionally independent. The subgraph probability of the three cliques is simply the product of their conditional subgraph probabilities. ■

We scan the cliques in a clique tree in the depth first order. When we reach a leaf clique C_l, its subgraph probability is calculated and sent to its parent clique C_p. If C_p only contains one child, then the subgraph probability of the subtree containing C_l and C_p is computed at C_p, using Theorem 7.4. If C_p contains more than one child, then the subgraph probability of all cliques in the subtree with root C_p is computed at C_p as described in Example 7.5. The complete procedure is shown in Algorithm 7.1.

Generally, if there are d chains of cliques connecting to a clique C, calculating the subgraph probability for each chain takes $O(n_i^2)$ where n_i is the number of cliques contained in the i-th chain. Calculating the subgraph probability for all cliques takes $O(dn^2)$, where $n = \sum_{1\leq i\leq d} n_i$ is the number of cliques in all d chains.

Algorithm 7.1 Computing subgraph probabilities.

Input: A dominant subgraph $G(v)$
Output: Subgraph probabilities $f(x) = Pr(G(v),x)$
Method:
1: **for all** connected component G_i in $G(v)$ $(1 \leq i \leq m)$ **do**
2: $f_i(x) = RecursiveSubgraph(G_i, G_i.root)$;
3: **end for**
4: $f(x) = f_i(x) * \cdots * f_m(x)$ {*Symbol $*$ denotes the convolution operation}

RecursiveSubgraph(G,C)
Input: Component G and clique C
Output: Subgraph probabilities $f_{G_C}(x)$ of G_C (the subtree with root C)
Method:
1: **for all** children C_i of C $(1 \leq i \leq m)$ **do**
2: $f_{G_{C_i}}(x) = RecursiveSubgraph(G_i, C_i)$;
3: **end for**
4: compute $f_0(x) = Pr(G_C, x | \neg v_1 \cdots \neg v_m)$;
5: **for** $i = 1$ to m **do**
6: compute $f_i(x) = Pr(G_C, x | v_i)$;
7: **end for**
8: return $f_{G_C}(x)$ by integrating $f_0(x), \cdots, f_m(x)$;

Computing subgraph probabilities when v is in $G(v)$

Let C be the clique containing v. When v appears, the other vertices in C cannot appear. By plugging in this constraint, the conditional subgraph probability $Pr(G(v),x|v)$ can be computed using the same method discussed in this section.

7.4.3 Integrating Multiple Components

Once the subgraph probability distribution of each connected component is obtained, we can calculate the convolution of those subgraph probabilities as the subgraph probability distribution over all connected component.

Theorem 7.5 (Integrating multiple components). *Given a dominant subgraph $G(v)$, let $G_1, \cdots, G_m$ be a set of connected components of $G(v)$. Let n_i be the number of cliques in G_i, then,*

1. $f_1(x) = Pr(G_1, x)$;
2. For $2 \leq i \leq m$, $f_i(x) = Pr(\bigcup_{j=1}^{i} G_j, x) = \sum_{x_i=0}^{n_i} f_{i-1}(x - x_i) Pr(G_i, x_i)$. ■

The complexity of computing the distribution $Pr(G(v),x)$ is $O(n^2)$, where n is the number of cliques in $G(v)$.

7.5 Exact Query Answering Algorithms

In this section, we first develop an exact query evaluation algorithm based on the subgraph probability calculation technique discussed in Section 7.3. Then, we discuss how to improve the efficiency of the algorithm.

7.5.1 An Exact Algorithm

With the subgraph probability computation technique discussed in Section 7.3, we can answer a probabilistic threshold top-k query on a set of probabilistic linkages as follows.

First, we build the corresponding PME-graph G and sort all vertices satisfying the query predicate P in the ranking order according to the scoring function f. Let $S = v_1, \cdots, v_n$ be the list of vertices in the ranking order. We then scan the vertices one by one.

For each vertex v_i, we derive the dominant subgraph $G(v_i)$ of v_i. The subgraph probability of $G(v_i)$ can be computed using the method discussed in Section 7.4. The top-k probability of v_i can be calculated using Equation 7.6.

7.5.2 Reusing Intermediate Results

Can we reuse the intermediate results to reduce the computational cost? Once the subgraph probability of a vertex is calculated, the intermediate results can be reused to compute the subgraph probability of the next vertex in the ranked list.

Example 7.6 (Reusing intermediate results). Consider the PME-graph G in Figure 7.7(a) that contains three connected components G_1, G_2 and G_3. Suppose the list of vertices $v_1, \cdots, v_{10}$ are sorted in the ranking order.

To compute the top-k probability of v_5, we first construct the dominant subgraph $G(v_5)$, which contains the grey vertices in Figure 7.7(a).

To compute the top-k probability of v_6, we construct the dominant subgraph $G(v_6)$ as shown in Figure 7.7(b).

Interestingly, by comparing $G(v_5)$ and $G(v_6)$, we can find the following.

First, in component G_1, the dominant graph of v_5 and v_6 are the same. Therefore, $Pr(G_1(v_6),x) = Pr(G_1(v_5),x)$.

Moreover, in component G_2, the subgraph probability of $G_2(v_5)$ is $Pr(G_2(v_5),x) = Pr(\{v_2\},x)$. Since $v_6 \in G_2$, we have to compute the conditional subgraph probability of $G_2(v_6)$, $Pr(G_2(v_6),x|v_6) = Pr(\{v_2\},x|v_6)$. Since $Pr(\{v_2\},x)$ has been computed when scanning v_5, we can compute $Pr(\{v_2\},x|v_6)$ based on $Pr(\{v_2\},x)$ using Theorem 7.4.

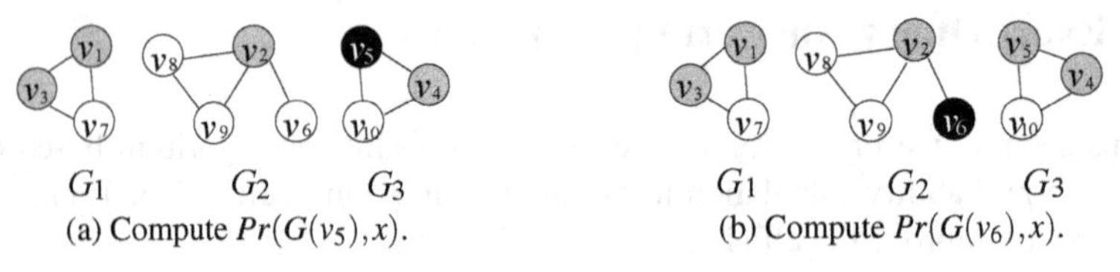

(a) Compute $Pr(G(v_5),x)$.

(b) Compute $Pr(G(v_6),x)$.

Fig. 7.7 Reuse the intermediate results.

Last, in component G_3, suppose we have computed the conditional subgraph probability $Pr(G_3(v_5),x|v_5) = Pr(\{v_4\},x|v_5)$. Then, we can compute the subgraph probability $Pr(G_3(v_6),x) = Pr(\{v_4,v_5\},x)$ based on $Pr(G_3(v_5),x|v_5)$ using Equation 7.7. ■

Generally, let $S = \{v_1,\cdots,v_n\}$ be the set of vertices satisfying P and sorted in the ranking order. Let $G(v_i)$ be the dominant graph of v_i. After obtaining the subgraph probability of $G(v_i)$, we scan v_{i+1} and compute the subgraph probability for $G(v_{i+1})$. For each component G_j, one of the following four cases happens.

Case 1: neither v_i nor v_{i+1} is in G_j.

Then, $G_j(v_{i+1}) = G_j(v_i)$. G_1 in Figure 7.7 illustrates this case.

Case 2: only v_i is in G_j.

G_3 in Figure 7.7 is an example of this case. After the conditional subgraph probability $Pr(G_j(v_i),x|v_i)$ has been computed, when scanning v_{i+1}, we want to compute the subgraph probability $Pr(G_j(v_{i+1}),x)$. The following corollary shows that $G_j(v_i)$ is a subgraph of $G_j(v_{i+1})$.

Corollary 7.3 (Property of dominant subgraphs).
Given a PME-graph $G = (V,E)$ and two vertices v_i and v_j, let $G(v_i)$ and $G(v_j)$ be the dominant subgraphs of v_i and v_j, respectively. If $v_i \preceq_f v_j$, then $G(v_i)$ is a subgraph of $G(v_j)$.

Proof. There are two categories of vertices in $G(v_i)$. Let v be a vertex in $G(v_i)$. We only need to show that v is also in $G(v_j)$.

If $v \preceq_f v_i$, then $v \preceq_f v_j$ (since $v_i \preceq v_j$). Thus, v is also in $G(v_j)$.

If v is not ranked higher than v_i, then v must be a joint vertex lying in a path $\langle v_1,\cdots,v,\cdots,v_2\rangle$ where $v_1 \preceq_f v_i$ and $v_2 \preceq_f v_i$. Since $v_i \preceq_f v_j$, we have $v_1 \preceq_f v_j$ and $v_2 \preceq_f v_j$. Thus, v is also in $G(v_j)$. ■

Therefore, we can compute the subgraph probability $Pr(G_j(v_{i+1}),x)$ based on $Pr(G_j(v_i),x|v_i)$ using Equation 7.7.

Algorithm 7.2 Answering PT-k queries on probabilistic linkages.

Input: A PME-graph G containing components $G_1, \cdots, G_m$ and a PT-k query Q
Output: Answer to Q
Method:

1: sort all vertices satisfying the query predicate in the ranking order;
   {*let $S = v_1, \cdots, v_n$ be the vertices in the ranking order}
2: **for** $i = 2$ to n **do**
3:   **for** $j = 1$ to m **do**
4:     **if** G_j does not contain v_{i-1} or v_i **then**
5:       $f_j(x) = Pr(G_j(v_{i-1}), x)$;
6:     **else if** G_j contains v_{i-1} **then**
7:       apply Equation 7.6 to compute $f_j(x) = Pr(G_j(v_{i-1}), x)$;
8:     **else**
9:       apply Theorem 7.4 to compute $f_j(x) = Pr(G_j(v_{i-1}), x|v_i)$;
10:     **end if**
11:     $Pr(G(v), x|v_i) = f(x) = f_1(x) * \cdots * f_m(x)$;
12:     $Pr^k(v_i) = Pr(v_i) \sum_{i=0}^{k-1} Pr(G(v), x|v_i)$;
13:   **end for**
14: **end for**

Case 3: only v_{i+1} is in G_j.

G_2 in Figure 7.7 illustrates this case. When scanning v_i, we computed the subgraph probability for $Pr(G_j(v_i), x)$. Then, when processing v_{i+1}, we have to compute $Pr(G_j(v_{i+1}), x|v_{i+1})$. Again, according to Corollary 7.3, $G_j(v_i)$ is a subgraph of $G_j(v_{i+1})$. Thus, $Pr(G_j(v_{i+1}), x|v_{i+1})$ can be computed based on $Pr(G_j(v_i), x)$ using Theorem 7.4.

Case 4: both v_i and v_{i+1} are in G_j.

The situation is similar to Case 3. Subgraph probability $Pr(G_j(v_{i+1}), x|v_{i+1})$ can be computed based on $Pr(G_j(v_i), x|v_i)$ using Theorem 7.4.

Generally, we scan each vertex v_i in the ranked list and compute the subgraph probabilities of $G(v_i)$ in each component G_j. If a component does not contain v_i or v_{i-1}, then the subgraph probabilities of v_i and v_{i-1} are the same. Otherwise, we apply Equation 7.7 or Theorem 7.4 to compute the subgraph probabilities, which avoids computing subgraph probabilities from scratch. Algorithm 7.2 shows the complete routine.

7.5.3 Pruning Techniques

From Corollary 7.3, for any two vertices $v_i \preceq_f v_j$, the dominant subgraph $G(v_i)$ is a subgraph of $G(v_j)$. Therefore, we have

$$\sum_{1 \le a \le x} Pr(G(v_i), a) \le \sum_{1 \le b \le x} Pr(G(v_j), b) \tag{7.8}$$

Theorem 7.6 (Pruning rule). Given a PME-graph G and a probabilistic threshold top-k query $Q^k_{P,f}$ with probability threshold $p \in (0,1]$, if for vertex v, $\sum_{1 \le a \le k-1} Pr(G(v_i), a) < p$, then for any vertex v' ($v \preceq_f v'$), $Pr^k(v') < p$. ■

Theorem 7.6 states that, once the subgraph probability for the dominant subgraph of a vertex v fails the probability threshold, all vertices ranked lower than v can be pruned.

7.6 Extensions to Aggregate Queries

Interestingly, the above techniques for ranking queries on uncertain data can be used to answer aggregate queries. In this section, we discuss the aggregate query answering on probabilistic linkages.

7.6.1 Aggregate Queries on Probabilistic Linkages

Example 7.7 (Aggregate queries). Consider Example 2.12 again. If a medical expert is interested in counting the number of linkages between the hospitalization registers and the causes-of-death registers. The answer to Q directly corresponds to the death population after hospitalization.

Suppose $\delta_M = 0.9$ and $\delta_U = 0.45$. No records in Table 2.4 are considered matched, since the linkage probabilities are all lower than δ_U. Is the answer to Q simply 0?

Record a_1 in the hospitalization register data set is linked to three records in the causes-of-death register data set, namely b_1, b_2 and b_3, with linkage probability 0.3, 0.3 and 0.4, respectively. Therefore, the probability that "John H. Smith" is linked to some records in the causes-of-death register data set and thus reported dead is $0.3 + 0.3 + 0.4 = 1$. Similarly, the probability that "Johnson R. Smith" is reported dead is 1. ■

In a possible world, the answer to an aggregate query is certain. Therefore, the answer to an aggregate query is a multiset of the answers in the possible worlds. Incorporating the probabilities, the answer to an aggregate query is a probability distribution on possible answers.

Definition 7.8 (Aggregate query on linkages). Given a set $\mathscr{L}_{A,B}$ of linkages between tables A and B, let Q^P_F be an **aggregate query**, where P is a predicate which may involve attributes in A, B, or both and F is an aggregate function on attribute $\mathscr{A}$. The **answer to Q^P_F on linkages** is the probability distribution

$f(v) = Pr(Q_F^P(\mathscr{L}_{A,B}) = v) = \sum_{W \in \mathscr{W}_{\mathscr{L},A,B}, Q_F^P(W)=v} Pr(W)$, where W is a possible world, $Q_F^P(W)$ is the answer to Q_F^P on the linkages in W, and $Pr(W)$ is the probability of W. ■

On a large set of linkages, there may be a huge number of possible worlds. Computing a probability distribution completely may be very costly. Moreover, if there are many possible answers in the possible worlds, enumerating all those values is hard to be manipulated by users. Since histograms are popularly adopted in data analytics and aggregate query answering, here we advocate answering aggregate queries on linkages using histograms. We consider both equi-width histograms and equi-depth histograms.

Definition 7.9 (Histogram answer). Consider an aggregate query Q on linkages $\mathscr{L}$, let v_{min} and v_{max} are the minimum and the maximum values of Q on all possible worlds.

Given a bucket width parameter η, and a minimum probability threshold τ, the **equi-width histogram answer** to Q is a set of interval tuples (ϕ_i, p_i) $(1 \leq i \leq \lceil \frac{v_{max}-v_{min}}{\eta} \rceil)$ where $\phi_j = [v_{min} + (j-1)\eta, v_{min} + j\eta)$ $(1 \leq j < \lceil \frac{v_{max}-v_{min}}{\eta} \rceil)$ and $\phi_{\lceil \frac{v_{max}-v_{min}}{\eta} \rceil} = [v_{min} + (\lceil \frac{v_{max}-v_{min}}{\eta} \rceil - 1)\eta, v_{max}]$ are $\lceil \frac{v_{max}-v_{min}}{\eta} \rceil$ equi-width intervals between v_{min} and v_{max} where $p_i = Pr(Q(\mathscr{L}) \in \phi)$. An interval pair (ϕ_i, p_i) is output only if $p_i \geq \tau$.

Given an integer $k > 0$, the **equi-depth histogram answer** to Q is a set of interval tuples (ϕ_i, p_i) $(1 \leq i \leq k)$ where $\phi_j = [v_{j-1}, v_j)$ $(1 \leq j < k$, $v_0 = v_{min}$, and $v_j = \arg\min_x \{x | Pr(Q(\mathscr{L}) \leq x) \geq \frac{j}{k}\})$ and $\phi_k = [v_{k-1}, v_{max}]$. ■

7.6.2 Count, Sum and Average Queries

To answer a `count` query on a set of linkages, we can apply the same technique discussed in Sections 7.3 and 7.4. Once the `count` probability distribution is calculated, we can compute the answer to the aggregate histogram query by partition the `count` values into η equi-width intervals or k equi-depth intervals.

The bottleneck of answering count queries are integrating multiple chains in a connected component and integration multiple components. We introduce two approximation techniques that accelerate the computation for equi-width histogram answer and equi-depth histogram answer, respectively.

7.6.2.1 Equi-width Histogram Answer Approximation

When computing the `count` probability for G using Theorem 7.5, intuitively, we can ignore the values whose probabilities are very small.

Let $x_1, \cdots, x_m$ be the list of values in $f_{i-1}(x)$ in the probability ascending order. Let $x_\mu = \max_{1 \leq j \leq m} \{x_j | \sum_{1 \leq h \leq j} Pr(x_h) < \varepsilon\}$, then we adjust the probabilities as

$$f'_{i-1}(x_i) = \begin{cases} 0, & 1 \le i \le \mu; \\ \frac{f_{i-1}(x_i)}{\sum_{\mu \le h \le m} Pr(x_h)}, & \mu < x \le m. \end{cases} \quad (7.9)$$

The computed answer is called the **ε-approximation** of the histogram answer. The quality of the approximation answer is guaranteed by the following theorem.

Theorem 7.7 (Approximation Quality). *Given a query Q on linkages with z components, a bucket width parameter η, and a minimum probability threshold τ, let (ϕ_i, p_i) be the equi-width histogram answer to Q, and $(\phi_i, \hat{p}_i)$ be the ε-approximation of (ϕ, p_i), then $|p_i - \hat{p}_i| < z\varepsilon$ for $1 \le i \le \lceil \frac{v_{max} - v_{min}}{\eta} \rceil$.*

Proof. We only need to show that each time when we integrate one connected component G_t $(2 \le t \le m)$, we introduce an approximation error of ε. Let $x_1, \cdots, x_m$ be the list of values in $f_{t-1}(x)$ in the probability ascending order, $v_1 = v_{min} + (i-1)\eta$ and $v_2 = v_{min} + i\eta$. Let $p^t_i = \sum_{v_1 \le x \le v_2} f_t(x)$ be the probability of bucket $[v_1, v_2)$. According to Theorem 7.5,

$$\begin{aligned} p^t_i &= \textstyle\sum_{v_1 \le x \le v_2} \sum_{b=1}^{m} f_{t-1}(x_b) Pr(Q(G_t) = x - x_b) \\ &= \textstyle\sum_{b=1}^{m} f_{t-1}(x_b) \sum_{v_1 - x_b \le x \le v_2 - x_b} Pr(Q(G_t) = x), \end{aligned}$$

where $f_{t-1}(x_b)$ is the exact count distribution in components $\{G_1, \cdots, G_{t-1}\}$. Let $\hat{p}^t_i = \sum_{v_1 \le x \le v_2} \hat{f}_t(x)$ be the approximate probability computed based on the ε-approximation $f'_{t-1}(x)$, then

$$\begin{aligned} \hat{p}^t_i &= \textstyle\sum_{v_1 \le x \le v_2} \sum_{b=1}^{m} f'_{t-1}(x_b) Pr(Q(G_t) = x - x_b) \\ &= \textstyle\sum_{b=1}^{m} f'_{t-1}(x_b) \sum_{v_1 - x_b \le x \le v_2 - x_b} Pr(Q(G_t) = x). \end{aligned}$$

Let $g(v_1 - x_b, v_2 - x_b)$ denote $\sum_{v_1 - x_b \le x \le v_2 - x_b} Pr(Q(G_t) = x)$. We have

$$\begin{aligned} p^t_i - \hat{p}^t_i &= \textstyle\sum_{b=1}^{m} f_{t-1}(x_b) g(v_1 - x_b, v_2 - x_b) - \sum_{b=1}^{m} f'_{t-1}(x_b) g(v_1 - x_b, v_2 - x_b) \\ &= \textstyle\sum_{d=1}^{\mu} f_{t-1}(x_d) g(v_1 - x_d, v_2 - x_d) \\ &+ \textstyle\sum_{b=\mu+1}^{m} \left(f_{t-1}(x_b) - f'_{t-1}(x_b)\right) \cdot g(v_1 - x_b, v_2 - x_b) \end{aligned} \quad (7.10)$$

Let $A = \sum_{d=1}^{\mu} f_{t-1}(x_d) g(v_1 - x_d, v_2 - x_d)$, then

$$\sum_{b=\mu+1}^{m} f_{t-1}(x_b) g(v_1 - x_b, v_2 - x_b) = p^t_i - A.$$

According to Equation 7.9, Equation 7.10 can be rewritten as

$$p^t_i - \hat{p}^t_i = A + \left(1 - \frac{1}{\sum_{\mu \le h \le m} f_{t-1}(x_h)}\right)(p^t_i - A)$$

On the one hand, since $\sum_{\mu \le h \le m} f_{t-1}(x_h) \le 1$ and $p^t_i > A$, we have $p^t_i - \hat{p}^t_i \le A$. Moreover, $A = \sum_{d=1}^{\mu} f_{t-1}(x_d) g(v_1 - x_d, v_2 - x_d) \le \sum_{d=1}^{\mu} f_{t-1}(x_d) \le \varepsilon$. Thus, $p^t_i - \hat{p}^t_i \le \varepsilon$.

On the other hand, $\sum_{\mu \leq h \leq m} f_{t-1}(x_h) \geq 1-\varepsilon$, and thus $1 - \frac{1}{\sum_{\mu \leq h \leq m} f_{t-1}(x_h)} \geq \frac{-\varepsilon}{1-\varepsilon}$. Moreover, $p_i^t - A \geq p_i^t - \varepsilon$. Therefore, $p_i^t - \hat{p}_i^t \geq -\varepsilon \times \frac{p_i^t - \varepsilon}{1-\varepsilon} \geq -\varepsilon$.

Therefore, after integrating m components, we have $|p_i - \hat{p}_i| \leq \varepsilon$. ■

After the probability adjustment, $f_{i-1}'(x) > \varepsilon$ holds for each value x with non-zero probability. Thus, the number of values with non-zero probability is at most $\frac{1}{\varepsilon}$. Therefore, the overall complexity of computing the ε-approximation of $f_i(x)$ is $O(\frac{1}{\varepsilon^2})$. The overall complexity is $O(\frac{m}{\varepsilon^2})$, where m is the number of components in G.

7.6.2.2 Equi-depth Histogram Answer Approximation

To accelerate probability calculation for equi-depth histogram answers, we introduce an approximation method that keeps a constant number of values in the intermediate results.

In theorem 7.5, if $f_{i-1}(x)$ contains values $x_1, \cdots, x_{n_{i-1}}$ $(n_{i-1} > k)$ in the value ascending order, then we compute the ρ-quantiles $x_i' = \arg\min_x \{Pr(Q(G) \leq x) \geq \frac{i}{\rho}\}$ $(0 \leq i \leq \rho)$. From the $\rho + 1$ values, we construct an approximation of $f_{i-1}(x)$ as:

$$f_{i-1}'(x) = \begin{cases} \frac{1}{\rho}, \ x = \frac{x_{i-1}' + x_i'}{2} \ (1 \leq i \leq \rho); \\ 0, \ \text{otherwise.} \end{cases} \quad (7.11)$$

Then, the convolution of $f_{i-1}'(x)$ and $Q(G_i)$ are used to estimate $f_i(x)$. The approximation quality is guaranteed by the following theorem.

Theorem 7.8 (Approximation quality). *Given a query Q on a PME-graph G with m components, an integer $k > 0$, let (ϕ_i, p_i), where $\phi_i = [v_{i-1}, v_i)$ $(1 \leq i \leq k)$, be the equi-depth histogram answer computed using the ρ-quantile approximation, then*

$$|Pr(Q(G) \leq v_i) - Pr'(Q(G) \leq v_i)| < \frac{m}{\rho}$$

where $Pr'(Q(G) \leq v_i)$ is the probability computed using Equation 7.11.

Proof. We only need to show that each time when we integrate one connected component G_t $(2 \leq t \leq m)$, we introduce an approximation error of $\frac{1}{\rho}$.

Let $g_t(v_i) = Pr(Q(G) \leq v_i) = \sum_{x \leq v_i} f_t(x)$ and $g_{t-1}(x_b) = \sum_{x \leq x_b} f_{t-1}(x)$. Then, according to Theorem 7.5,

$$g_t(v_i) = \sum_{x_d=0}^{x_n} Pr(G_t, x_d) \cdot g_{t-1}(v_i - x_d)$$

where x_n is the number of cliques in G_t.

Let $g_{t-1}'(x_b) = Pr'(Q(G) \leq v_i) = \sum_{x \leq x_b} f_{t-1}'(x)$ where $f_{t-1}'(x)$ is the approximation of $f_{t-1}(x)$ using Equation 7.11, then

$$g'_t(v_i) = \sum_{x_d=0}^{x_n} Pr(G_t, x_d) \cdot g'_{t-1}(v_i - x_d)$$

Therefore,

$$|g_t(v_i) - g'_t(v_i)| = |\sum_{x_d=0}^{x_n} Pr(G_t, x_d) \cdot (g_{t-1}(v_i - x_d) - g'_{t-1}(v_i - x_d))|$$

Let x'_c $(1 \leq c \leq \rho)$ be the ρ-quantiles of $f_{t-1}(x)$. Suppose $x'_{c-1} \leq v_i - x_d \leq x'_c$ $(1 \leq c \leq \rho)$, there are two cases:

First, if $v_i - x_d \leq \frac{x'_{c-1}+x'_c}{2}$, then $g'_{t-1}(v_i - x_d) = \frac{c-1}{\rho}$ and $\frac{c-1}{\rho} \leq g_{t-1}(v_i - x_d) \leq \frac{c}{\rho}$. Thus, $0 \leq g_{t-1}(v_i - x_d) - g'_{t-1}(v_i - x_d) \leq \frac{1}{\rho}$.

Second, if $v_i - x_d > \frac{x'_{c-1}+x'_c}{2}$, then $g'_{t-1}(v_i - x_d) = \frac{c}{\rho}$ and $\frac{c-1}{\rho} \leq g_{t-1}(v_i - x_d) \leq \frac{c}{\rho}$. Thus, $-\frac{1}{\rho} \leq g_{t-1}(v_i - x_d) - g'_{t-1}(v_i - x_d) \leq 0$.

In both cases, $|g_{t-1}(v_i - x_d) - g'_{t-1}(v_i - x_d)| \leq \frac{1}{\rho}$. Therefore,

$$|g_t(v_i) - g'_t(v_i)| \leq \sum_{x_d=0}^{x_n} Pr(G_t, x_d) \cdot \frac{1}{\rho} = \frac{1}{\rho}.$$

■

Using the above approximation method, the overall complexity of computing the approximate k equi-depth histogram answer is $O(m\rho^2)$, where m is the number of connected components.

7.6.2.3 Sum/Average Queries

The query answering algorithm for `count` queries can be extended to evaluate `sum` and `average` queries with minor changes.

To answer a `sum` query $Q^P_{\texttt{sum}}$, we apply the same preprocessing techniques discussed in Sections 7.3 and 7.4. The only difference is that the private vertices in the same clique with the same value in attribute $\mathscr{A}$ are compressed as one vertex. When scanning each clique, the `sum` distribution of its subtree are computed and passed to its parent clique. The overall complexity is $O(n^2)$, where n is the number of values in v_{min}, the smallest value of any linkage in attribute $\mathscr{A}$, and v_{max}, the sum of the values of all linkages in attribute $\mathscr{A}$. The `average` query can be evaluated similarly.

Name	Age
Larry Stonebraker	35
Richard Ramos	45
Catherine Spicer	46
Bruce Mourer	47
Jason Haddad	51
Angelina Amin	53
Jo Avery	53
Nicola Stewart	54
Alvin Wood	54
Gerald McMullen	55

Table 7.1 The top-10 youngest patients in the cancer registry.

7.6.3 Min and Max Queries

A `min` query $Q^P_{\texttt{min}}$ is a special top-k query with $k = 1$. For a vertex v_i whose value in attribute $\mathscr{A}$ is x_i. The `min` probability of x_i is the probability that v_i is ranked at the first place in possible worlds. Therefore, the query answering techniques discussed in Sections 7.3 and 7.4 can be directly applied to answering `min` queries. The only difference is that we want to derive the histogram answers to a `min` query.

The `min` probability of $x_i = v_i.\mathscr{A}$ can be bounded as

$$Pr(Q(G) = x_i) \leq \prod_{G_j \subseteq G, v_i \notin G_j} Pr(\wedge_{v \in G_j \cap S_{i-1}} \neg v) \tag{7.12}$$

where $G_j \subseteq G$ and $G_j \cap S_{i-1} \neq \emptyset$.

Let $\pi(x_i) = \prod_{G_j \subseteq G, v_i \notin G_j} Pr(\wedge_{v \in G_i \cap S_{i-1}} \neg v)$, we can derive the following pruning rule.

Theorem 7.9 (Pruning rule). *Given a PME-graph G, a* `min` *query Q on attribute* $\mathscr{A}$*, a bucket width parameter* η*, and a minimum probability threshold* τ*, let S be the ranked list of linkages on attribute* $\mathscr{A}$ *and* $x_i = v_i.\mathscr{A}$ *be the value of the i-th vertex in S, if* $\pi(x_i) < \frac{\tau}{\eta}$*, then for any interval tuple* (ϕ_j, p_j) *where* $\phi_j = [v_{min} + (j-1) \cdot \eta, v_{min} + j \cdot \eta]$ *and* $v_{min} + (j-1) \cdot \eta \leq x_i$*,* $p_j < \tau$*.*

Proof. The theorem follows with Equation 7.12 directly. ∎

Theorem 7.9 states that, for a tuple interval ϕ_j, once the smallest value in the interval ϕ_j is smaller or equal to x_i, its `min` probability is smaller than $\frac{\tau}{\eta}$. Thus, the interval tuple is not output.

To compute the answer distribution of a maximum query $Q^P_{\texttt{max}}$, we can apply the same methods as for processing the minimum query. The only difference is, the linkages are ranked in the descending order of their values in attribute $\mathscr{A}$. The details are omitted for the interest of space.

The complexity of computing the probability of x_i being the minimal value is $O(n^2)$ where n is the number of cliques in the PME-graph.

Name	Age	Top-10 probability
Larry Stonebraker	35	0.8375
Catherine Spicer	46	0.775
Bruce Mourer	47	0.87875
Jason Haddad	51	0.85625
Angelina Amin	53	0.885
Jo Avery	53	0.7975
Nicola Stewart	54	0.8575
Tiffany Marshall	57	0.86
Bridget Hiser	58	0.778

Table 7.2 The patients in the cancer registry whose top-10 probabilities are at least 0.3.

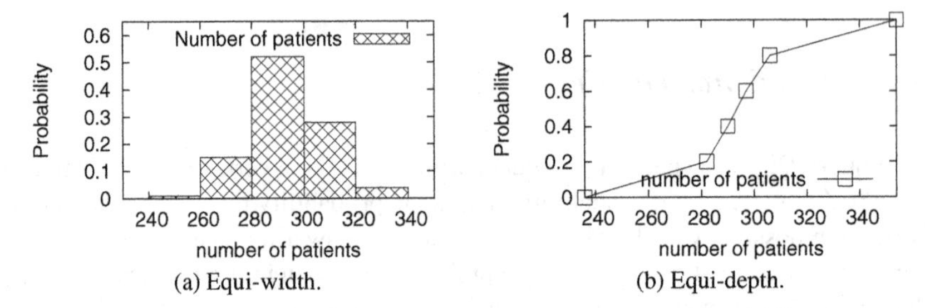

(a) Equi-width. (b) Equi-depth.

Fig. 7.8 Answer to query: the number of patients appearing in both data sets.

7.7 Empirical Evaluation

In this section, we report a systematic empirical study. All experiments were conducted on a PC computer with a 3.0 GHz Pentium 4 CPU, 1.0 GB main memory, and a 160 GB hard disk, running the Microsoft Windows XP Professional Edition operating system, Our algorithms were implemented in Microsoft Visual Studio 2005.

7.7.1 Results on Real Data Sets

First, we apply the ranking queries and aggregate queries on the Cancer Registry data set and the Social Security Death Index provided in Link Plus 2.0[1].

The Cancer Registry data set contains 50,000 records and each record describes the personal information of a patient, such as `name` and `SSN`. The Social Security Death Index data set contains 10,000 records and each record contains the personal information of an individual, such as `name`,`SSN` and `Death Date`. Since the information of some records are incomplete or ambiguous, we cannot find the exact match for records in the two data sets.

[1] `http://www.cdc.gov/cancer/npcr/tools/registryplus/lp.htm`

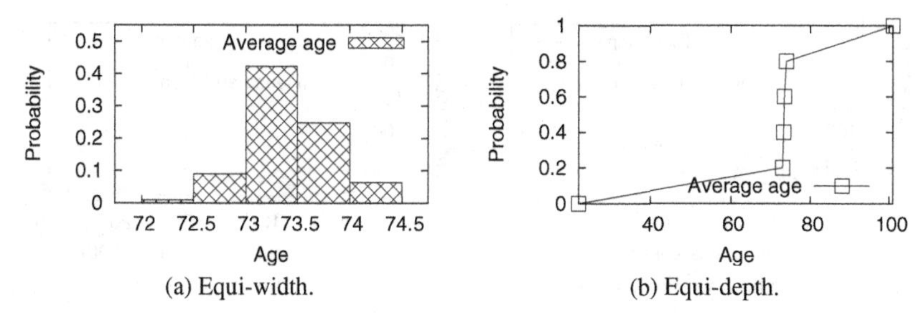

(a) Equi-width. (b) Equi-depth.

Fig. 7.9 Answer to query: the average age of the patients appearing in both data sets.

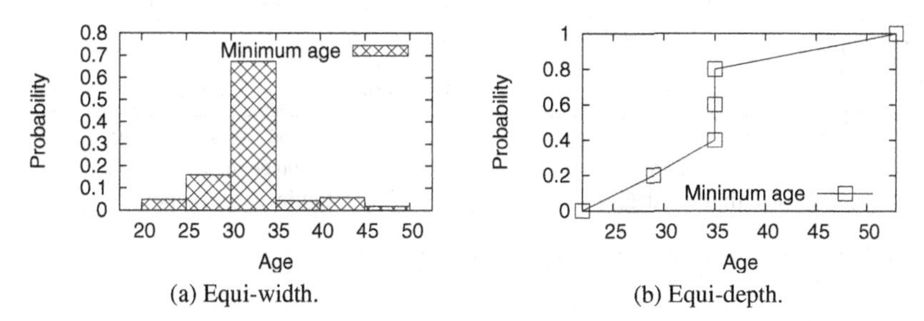

(a) Equi-width. (b) Equi-depth.

Fig. 7.10 Answer to query: the minimal age of the patients appearing in both data sets.

Link Plus is a popularly used tool that computes the probability that two records referring to the same individual. It matches the records on the two data sets based on `name`, `SSN` and `Date of Birth` and returns 4,658 pairs of records whose linkage probabilities are greater than 0. The system suggests that a user should set a matching linkage probability threshold. The pairs of records passing the threshold are considered matching. If we set the threshold as the default value 0.25 suggested by the system, only 99 pairs of records are returned.

First, we want to find the top-10 youngest patients in the cancer registry and reported death. Therefore, we ask a probabilistic top-k query with $k = 10$ and $p = 0.3$. For each linked pair, we use the average ages in the Cancer Registry and the Social Security Death Index. If we only consider the linked pairs whose probability pass the matching threshold 0.25, then the top-10 youngest patients with their edges are shown in Table 7.1. However, we consider all linked pairs whose matching probabilities are greater than 0 and find the patients whose top-10 probability is greater than 0.3, we can the results as shown in Table 7.2.

Then, we ask the following `count` query on the data sets: *what is the number of patients appearing in both data sets?* The histogram answers are shown in Figure 7.8 It is far from the 99 returned on the linked pairs passing the matching threshold.

Moreover, an `average` query *finds out the average age of the patients appearing in both data sets*. If only the 99 records whose matching probabilities are above

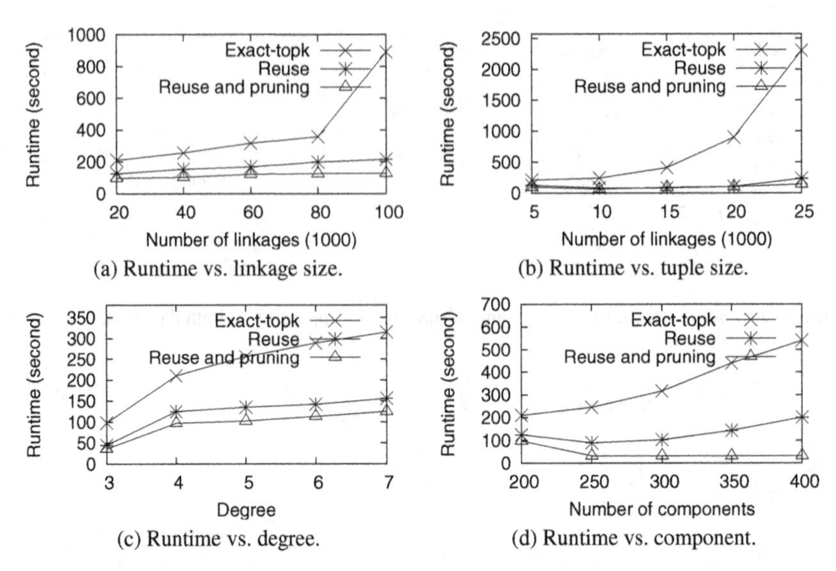

(a) Runtime vs. linkage size. (b) Runtime vs. tuple size.

(c) Runtime vs. degree. (d) Runtime vs. component.

Fig. 7.11 Efficiency and scalability of PTK query evaluation.

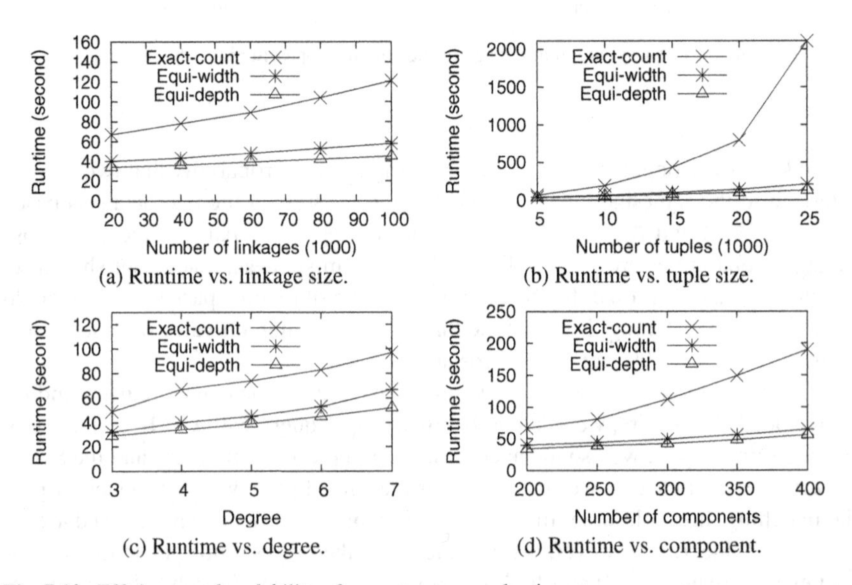

(a) Runtime vs. linkage size. (b) Runtime vs. tuple size.

(c) Runtime vs. degree. (d) Runtime vs. component.

Fig. 7.12 Efficiency and scalability of `count` query evaluation.

0.25 are considered, the `average age` is 71.7. However, if we compute the probability distribution of the average age over all linked pairs, the answer is very different. The histogram answers are shown in Figure 7.9.

Last, it is also interesting to know the smallest age of the patients who are in the cancer registry and reported death. Therefore, we ask a `min` query : *what is the*

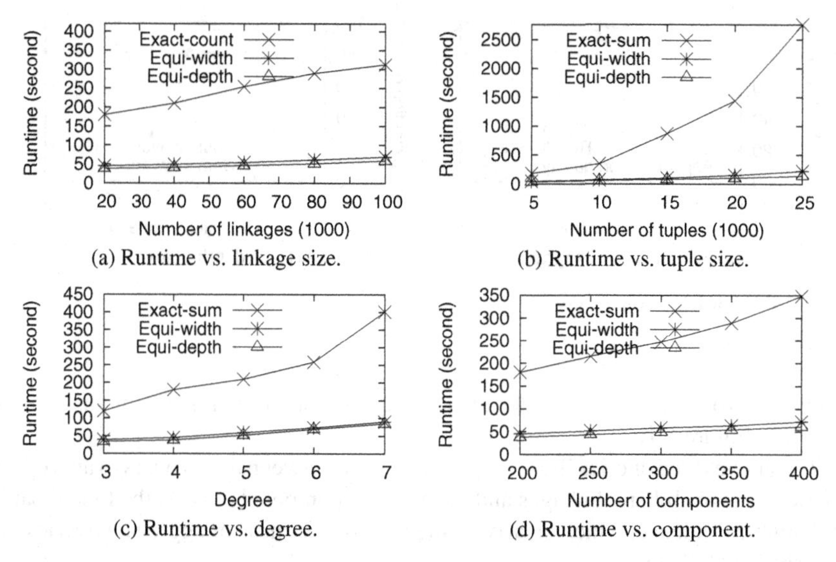

(a) Runtime vs. linkage size. (b) Runtime vs. tuple size.

(c) Runtime vs. degree. (d) Runtime vs. component.

Fig. 7.13 Efficiency and scalability of `sum` query evaluation.

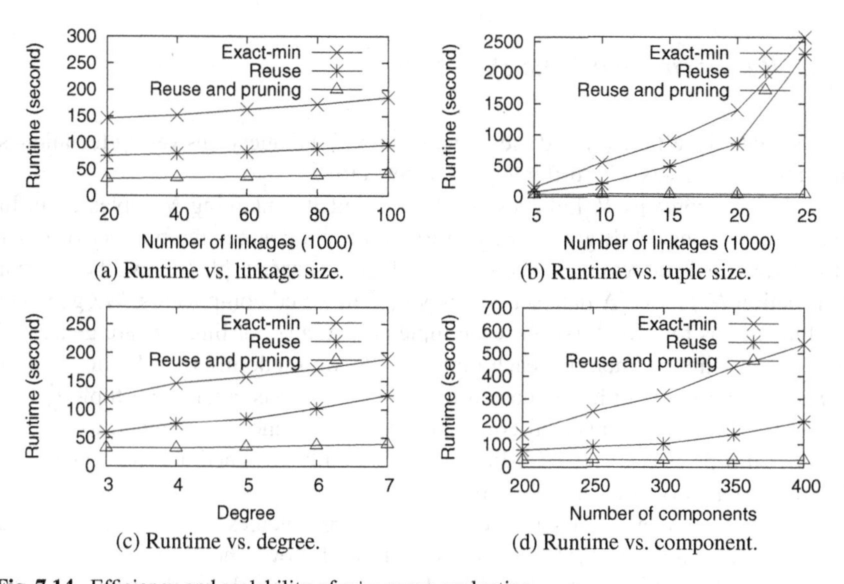

(a) Runtime vs. linkage size. (b) Runtime vs. tuple size.

(c) Runtime vs. degree. (d) Runtime vs. component.

Fig. 7.14 Efficiency and scalability of `min` query evaluation.

minimal age of the patients appearing in both data sets? For each linked pair, we use the average ages in the Cancer Registry and the Social Security Death Index. If we only consider the linked pairs whose probability pass the matching threshold 0.25, then the minimum age is 35. However, by considering all linked pairs whose

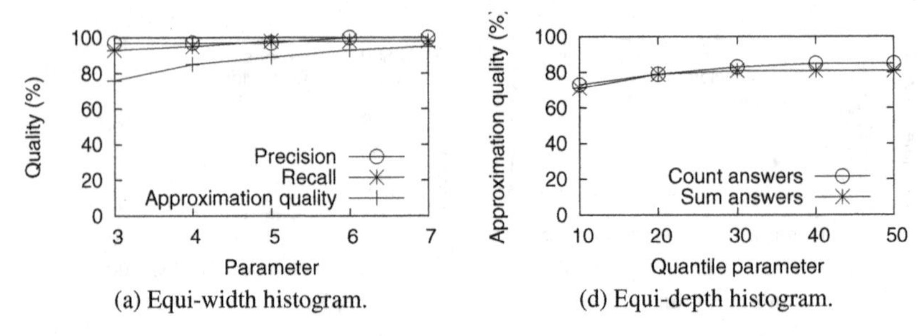

(a) Equi-width histogram. (d) Equi-depth histogram.

Fig. 7.15 Approximation quality.

matching probabilities are greater than 0, we can obtain the histogram answers as shown in Figure 7.10.

The above three queries illustrate the difference between the answers to aggregate queries on probabilistic linkages and the answers computed only on the linked pairs with high confidence, which clearly shows the effectiveness of aggregate queries on probabilistic linkages.

7.7.2 Results on Synthetic Data Sets

We evaluate the efficiency and the effectiveness of the query answering techniques in synthetic data sets with different parameter settings.

A data set contains N_l linkages in tables A and B containing N_t tuples each. In the corresponding PME-graph, the degree of a tuple denotes the number of other tuples (maximal cliques) it connects to. The degree of each tuple follows the Normal distribution $N(\mu_t, \sigma_t)$. A data set contains N_c connected components. We generate the linkages as follows. First, for each tuple $t_A \in A$, a set of linkages are generated associating with t_A. Then, for each tuple $t_B \in B$, we randomly pair the tuples in A to t_B. In order to avoid loops, once a linkage $\mathscr{L}(t_A, t_B)$ is created, all tuple $t'_A \in A$ that are in the same connected component with t_A cannot be assigned to t_B. The membership probability of each linkage is randomly assigned and normalized so that the probability of each tuple is between $(0, 1]$.

By default, a synthetic data set contains 20,000 linkages between tables A and B with 5,000 tuples each. The degree of a tuple follows the Normal distribution $N(4, 1)$. The bucket width $\eta = 1000$ and the minimum probability threshold $\tau = 0.1$. The parameter k for equi-depth histogram answer is set to 10. We set parameter $k = 1000$ and $p = 0.4$.

First, we evaluate the efficiency and scalability of the query answering methods. All algorithm developed are scalable when the number of linkages is as large as 100,000.

Figures 7.11(a) to (d) show the runtime of the query answering methods for probabilistic threshold top-k queries in various parameter settings. *Exact-topk* is described in Algorithm 7.1. *Reuse* denotes the algorithm with the reusing technique and *Reuse+pruning* denotes the algorithm with both the reusing technique and the pruning technique.

Figures 7.12(a) to (d) show the runtime of the query answering methods for `count` queries in various parameter settings. *Exact-count* is the exact algorithm. *Equi-width* and *Equi-depth* denotes the algorithms using the approximation techniques discussed in Sections 7.6.2.1 and 7.6.2.2, respectively. By default, $\varepsilon = 10^{-4}$ and $\rho = 30$.

In Figure 7.12(a), we fix the number of tuples to $5,000$ and the expected degree of each tuple is 4. The runtime of all three algorithms increase mildly as the number of linkage increases from $20,000$ to $100,000$, thanks to the vertex compression technique discussed in Section 7.3.3.

In Figure 7.12(b), we fix the expected degree of each tuple to 4 and increase the number of tuples in the data set. The runtime of the *exact-count* algorithm increases quadratically. The runtime of the two approximation algorithms increases very slightly.

Figure 7.12(c) shows the increase of runtime with respect to the degrees of tuples. The larger the degree, the more times of convolution are performed in the query evaluation. Therefore, the runtime of all three algorithms increases linearly with respect to the expected degree.

Last, Figure 7.12(d) illustrates the change of runtime when the number of connected components increases.

Figures 7.13(a) to (d) show the runtime of the `sum` query evaluation in the same setting as in Figure 7.12. The results on the `sum` query evaluation demonstrate the similar patterns as in the `count` query evaluation. But the overall cost for the `sum` query evaluation is higher than that of the `count` query evaluation, since it depends on the number of values appearing in the result distribution.

Figures 7.14(a) to (d) show the efficiency of the `min` query evaluation. *Exact-min* denotes the algorithm that transforms the `min` query to a set of `count` queries. *Reuse* denotes the algorithm that explores the sharing of computation among different linkages. *Reuse+Pruning* denotes the algorithm that applies the pruning technique discussed in Section7.5.3 in addition to the reuse method. It is shown that the two techniques improve the efficiency significantly.

Last, we evaluate the quality of the two approximation algorithms discussed in Sections 7.6.2.1 and 7.6.2.2, respectively.

The quality of ε-approximate euqi-width histogram answers are computed as $\frac{1}{|\{\phi | f(\phi) > \tau\}|} \sum_{f(\phi) > \tau} 1 - \frac{|\widehat{f}(\phi) - f(\phi)|}{f(\phi)}$, where $f(\phi)$ is the probability of interval ϕ and $\widehat{f}(\phi)$ is the probability of $x \in \phi$ estimated by the approximation method. The results are shown in Figure 7.15(a), with respect to different ε values 10^{-x} $(x = 3,4,5,6,7)$. The precision and recall are also plotted.

Figure 7.15(b) illustrates the approximation quality of the answers to both `count` and `sum` queries using the ρ quantiles. The quality is measured using $\frac{1}{k}\sum_{i=1}^{k} 1 - \frac{|\hat{f}(v_i) - f(v_i)|}{f(v_i)}$, where v_i is the value output as the approximation of the i-

th k-quantile, $\hat{f}(v_i)$ is probability computed using the approximation method and $f(v_i)$ is the real probability of v_i.

7.8 Summary

In this chapter, we investigate ranking queries evaluation on probabilistic linkages. In contrast to the traditional methods that use simple probability thresholds to obtain a set of deterministic linkages, we fully utilize the probabilities produced by the record linkage methods and consider the linked records as a distribution over possible worlds. By preserving the distribution information, we can provide more meaningful answers to aggregate queries, as illustrated in Section 7.7.1.

Moreover, we extend the ranking query evaluation method to answer aggregate queries. Efficient approximation and pruning techniques are developed.

Chapter 8
Probabilistic Path Queries on Road Networks

Path queries such as "find the shortest path in travel time from my hotel to the airport" are heavily used in many applications of road networks. Currently, simple statistic aggregates such as the average travel time between two vertices are often used to answer path queries. However, such simple aggregates often cannot capture the uncertainty inherent in traffic.

To capture the uncertainty in traffic such as the travel time between two vertices, in Section 2.3.3, we modeled the weight of an edge as an uncertain object that contains a set of samples. Moreover, we proposed three novel types of probabilistic path queries.

- A *probabilistic path query* asks a question such as "what are the paths from my hotel to the airport whose travel time is at most 30 minutes with a probability of at least 90%?"
- A *weight-threshold top-k path query* asks a question like "what are the top-3 paths from my hotel to the airport with the highest probabilities to take at most 30 minutes?"
- A *probability-threshold top-k path query* asks a question like "in terms of the travel time of a path guaranteed by a probability of at least 90%, what are the top-3 shortest paths from my hotel to the airport?"

To evaluate probabilistic path queries efficiently, in this chapter, we first develop three efficient probability calculation methods: an exact algorithm, a constant factor approximation method and a sampling based approach. Moreover, we devise the P* algorithm, a best-first search method based on a novel hierarchical partition tree index and three effective heuristic estimate functions. An extensive empirical study using real road networks and synthetic data sets shows the effectiveness of the proposed path queries and the efficiency of the query evaluation methods.

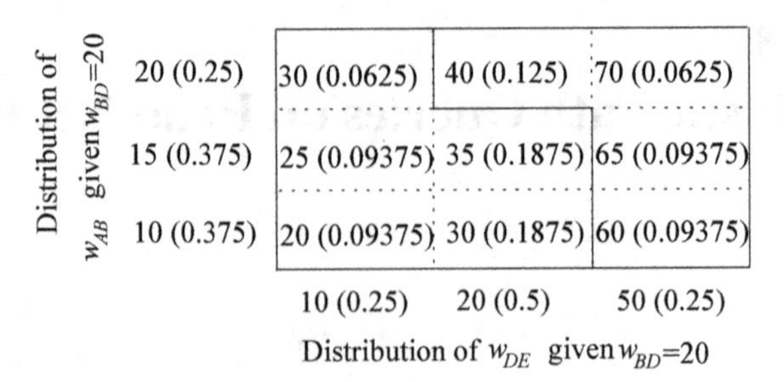

Distribution of
w_{AB} given w_{BD}=20
20 (0.25)
15 (0.375)
10 (0.375)
30 (0.0625)
40 (0.125)
70 (0.0625)
25 (0.09375)
35 (0.1875)
65 (0.09375)
20 (0.09375)
30 (0.1875)
60 (0.09375)
10 (0.25)
20 (0.5)
50 (0.25)
Distribution of w_{DE} given w_{BD}=20

Fig. 8.1 The weight constrained region of $P_1 = \langle A,B,D,E \rangle$ when w_{BD} takes sample 20.

8.1 Probability Calculation

There are two orthogonal issues in answering a path query $Q_l^\tau(u,v)$: the l-weight probability calculation and the path search. In this section, we first discuss how to compute the exact l-weight probabilities for paths. Then, two approximate algorithms are presented. We also present a straightforward depth-first path search method. An efficient best-first path search algorithm will be introduced in Section 8.2.

8.1.1 Exact *l*-Weight Probability Calculation

How can we calculate the l-weight probability of a path P when the edge weights are correlated? We can partition P into subpaths such that they are conditionally independent, as illustrated in the following example.

Example 8.1 (l-weight constrained region). Let $l = 55$ and $\tau = 0.5$. Consider path query $Q_l^\tau(A,E)$ and path $P = \langle A,B,D,E \rangle$ in the probabilistic graph in Figure 2.6.

P contains three edges $e_1 = (A,B)$, $e_3 = (B,D)$ and $e_6 = (D,E)$. The joint probabilities f_{e_1,e_3} and f_{e_3,e_6} are given in Figures 2.6(c) and 2.6(d), respectively, which specify the correlation between edges. Weights w_{e_1} and w_{e_6} are conditionally independent given w_{e_3}.

The conditional probability of w_{e_1} given $w_{e_3} = 20$ is

$$f_{e_1|e_3}(x|20) = \frac{f_{e_1,e_3}(x,20)}{f_{e_3}(20)} = \begin{cases} 0.375, & x = 10 \text{ or } x = 15; \\ 0.25, & x = 20. \end{cases}$$

The conditional probability of w_{e_6} given $w_{e_3} = 20$ is

$$f_{e_6|e_3}(x|20) = \frac{f_{e_6,e_3}(x,20)}{f_{e_3}(20)} = \begin{cases} 0.25, & x = 10 \text{ or } x = 50; \\ 0.5, & x = 20. \end{cases}$$

The probability that w_P is at most l when $w_{e_3} = 20$ is

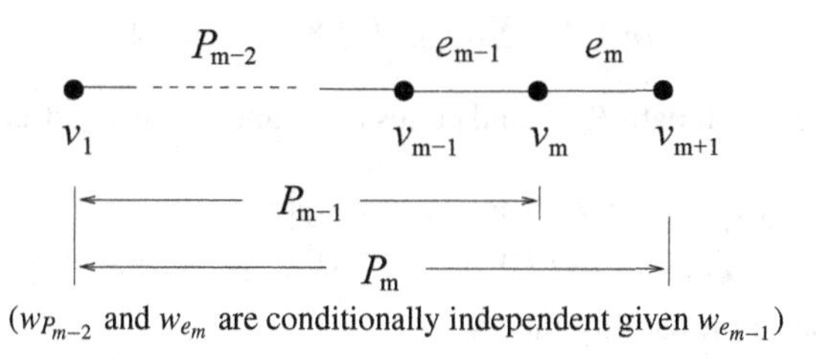

Fig. 8.2 A path P_m.

$$
\begin{aligned}
& Pr[w_P \leq 55 | w_{e_3} = 20] \\
= & \sum_{x_1+x_3 \leq 35} Pr[w_{e_1} = x_1, w_{e_6} = x_3 | w_{e_3} = 20] \\
= & \sum_{x_1+x_3 \leq 35} f_{e_1|e_3}(x_1|20) \cdot f_{e_6|e_3}(x_3|20)
\end{aligned}
$$

The sets of samples and their membership probabilities of $f_{e_1|e_3}(x_1|20)$ and $f_{e_6|e_3}(x_3|20)$ are $\{10(0.375), 15(0.375), 20(0.25)\}$ and $\{20(0.25), 20(0.5), 50(0.25)\}$, respectively. We sort the samples in the weight ascending order.

There are in total $3 \times 3 = 9$ samples on $w_{e_1} \times w_{e_6}$ when $w_{e_3} = 20$. To enumerate all samples, we can build a 3×3 sample array M as shown in Figure 8.1. Cell $M[i, j]$ stores two pieces of information: (1) $M[i, j].ws$ is the sum of the i-th sample of w_{e_1} and the j-th sample of w_{e_6}; and (2) $M[i, j].pr$ is the membership probability of sample $M[i, j].ws$ which equals the product of the corresponding membership probabilities. For example, the lowest left-most cell $M[1,1]$ corresponds to a sample where w_{e_1} is 10 and w_{e_6} is 10. Thus, $M[1,1].ws = 10 + 10 = 20$ and $M[1,1].pr = 0.375 \times 0.25 = 0.09375$.

*When w_{e_3} takes sample 20, in order to satisfy $w_{P_1} \leq 55$, the samples of $w_{e_1} + w_{e_6}$ should take values of at most 35. Those samples are at the lower-left part of the array. We call the region of those cells the l-*weight constrained region *when $w_{e_3} = 20$. The sum of the membership probabilities of the cells in the region is 0.625.*

The l-weight constrained region when w_{e_3} takes other samples can be calculated similarly. When $w_{e_3} = 25$ and $w_{e_3} = 30$, the sum of the membership probabilities of the cells in the l-weight constraint regions are 0.53125 and 0, respectively. $F_{P_1}(55) = f_{e_3}(20) \times 0.625 + f_{e_3}(25) \times 0.53125 + f_{e_3}(30) \times 0 = 0.4625 < \tau$. P_1 is not an answer to Q_l^τ. ■

The idea in Example 8.1 can be generalized to paths of arbitrary length.

Theorem 8.1 (Path weight distribution). *Let $P_m = \langle v_1, \ldots, v_{m+1} \rangle$, $P_{m-1} = \langle v_1, \ldots, v_m \rangle$ and $e_m = (v_m, v_{m+1})$ ($m > 2$), the conditional probability of w_{P_m} given w_{e_m} is*

$$
f_{P_m|e_m}(x|y) = \sum_{z \leq x-y} f_{e_{m-1}|e_m}(z|y) \times f_{P_{m-1}|e_{m-1}}(x-y|z) \tag{8.1}
$$

Moreover, the probability mass function of w_{P_m} is

$$f_{P_m}(x) = \Sigma_{y \leq x} f_{e_m}(y) \times f_{P_m|e_m}(x|y) \tag{8.2}$$

Proof. P_m contains subpath P_{m-2} and edges e_{m-1} and e_m, as illustrated in Figure 8.2. Therefore,

$$f_{P_m|e_m}(x|y) = Pr[w_{P_{m-1}} = x - y | w_{e_m} = y]$$
$$= \Sigma_{z_1 + z_2 = x - y} Pr[w_{P_{m-2}} = z_1, w_{e_{m-1}} = z_2 | w_{e_m} = y]$$

Using the basic probability theory,

$$Pr[w_{P_{m-2}} = z_1, w_{e_{m-1}} = z_2 | w_{e_m} = y]$$
$$= Pr[w_{e_{m-1}} = z_2 | w_{e_m} = y] Pr[w_{P_{m-2}} = z_1 | w_{e_{m-1}} = z_2]$$

Since $z_1 + z_2 = x - y$, we have

$$Pr[w_{P_{m-2}} = z_1 | w_{e_{m-1}} = z_2] = Pr[w_{P_{m-1}} = x - y | w_{e_{m-1}} = z_2]$$

Thus, Equation 8.1 holds. Equation 8.2 follows with the basic principles of probability theory. ■

For a path $P_m = \langle v_1, \ldots, v_{m+1} \rangle$, the probability function of w_{P_m} can be calculated from $w_{P_{m-1}}$ and w_{e_m} using Theorem 8.1. Calculating the probability mass function of P_m requires $O(|w_{P_{m-1}}| \cdot |w_{e_m}|) = O(\prod_{e \in P_m} |w_e|)$ time.

Interestingly, if only $F_{P_m}(l)$ is required, we do not need to use all samples in an edge weight. The largest sample we can take from w_{e_i} $(1 \leq i \leq m)$ is bounded by the following rule.

Lemma 8.1 (Sample components). *For weight threshold l and path $P = \langle v_1, \ldots, v_{m+1} \rangle$, a sample x_{i_j} of edge $e_i = (v_i, v_{i+1})$ $(1 \leq i \leq m)$ can be a component of a sample of P in the l-weight constrained region only if $x_{i_j} \leq l - \Sigma_{i' \neq i} x_{i'_1}$, where $x_{i'_1}$ is the smallest sample of edge $e_{i'} = (v_{i'}, v_{i'+1})$.*

Proof. Since one and only one sample should be taken from each edge in a possible world, the sum of the minimal sample from each edge in P is the smallest weight of P. Thus, the conclusion holds. ■

8.1.2 Approximating l-Weight Probabilities

The probability mass function of $F_{P_{m+1}}(l)$ can be calculated from the distributions on $f_{P_m|e_m}$ and $f_{e_m}|f_{e_{m+1}}$ according to Theorem 8.1. To accelerate probability calculation, we introduce an approximation method that keeps a constant number of samples in the weight distribution of any subpath during the search.

If w_{P_m} contains $n > 2t$ samples $x_1, \cdots, x_n$ (t is a user defined parameter), then we divide those values into b exclusive buckets $\phi_i = [x_{z_i}, x_{z'_i}]$, where

$$z_1 = 1, z_k = z'_{k-1} + 1, \text{ for } 1 < k \leq b;$$
$$z'_i = \max_{j \geq z_i} \{ j | F_{P_m|e_m}(x_j|y) - F_{P_m|e_m}(x_{z_i}|y) \leq \frac{1}{t} \}, \text{ for } 1 \leq i \leq b \tag{8.3}$$

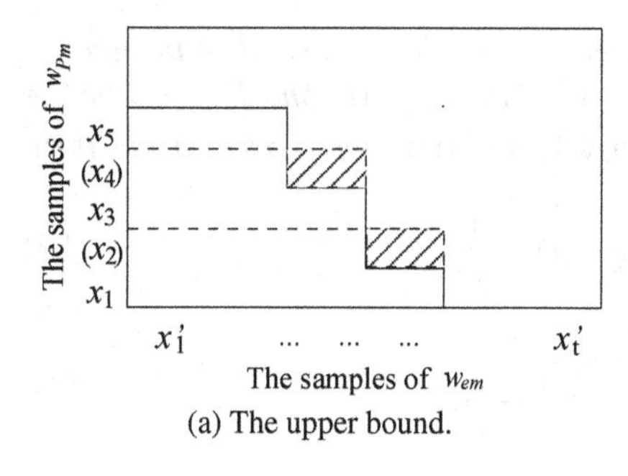

(a) The upper bound.

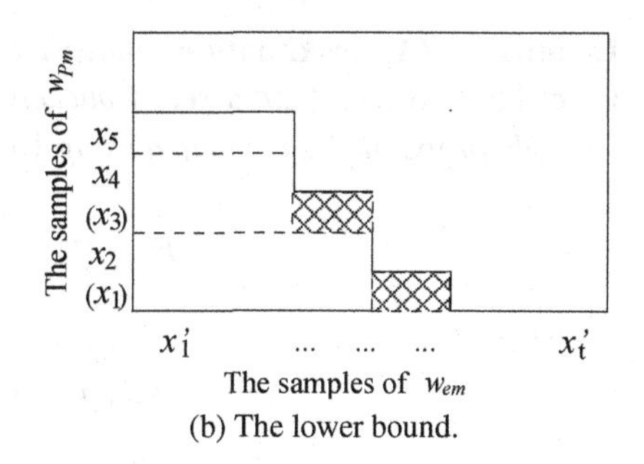

(b) The lower bound.

Fig. 8.3 The upper/lower bound of $F_{P_{m+1}}(l)$.

The number of buckets is at most $2t$, as shown in the following lemma.

Lemma 8.2 (Number of buckets). *Given w_{P_m} with n samples ($n > 2t > 0$), let $\phi_i = [x_{z_i}, x_{z'_i}]$ ($1 \leq i \leq b$) be b exclusive buckets satisfying Equation 8.3, then $b \leq 2t$.*
Proof. In the worst case, each bucket only contains one value in w_{P_m}, which means that the probability sum of any two consecutive values in w_{P_m} is greater than $\frac{1}{t}$. Then, if the number of values in w_{P_m} is greater than $2t$, the probability sum of all values in w_{P_m} will be greater than 1, which conflicts with the fact that w_{P_m} is a discrete random variable. ■

Constructing the buckets only requires one scan of all values in w_{P_m}. The minimal value in bucket $\phi_i = [x_{z_i}, x_{z'_i}]$ is $\min(\phi_i) = x_{z_i}$, and the maximal value in ϕ_i is $\max(\phi_i) = x_{z'_i}$. When computing the probability distribution of $w_{P_{m+1}}$ using w_{P_m}, we only select one value in each bucket $\phi_i \subset w_{P_m}$ as a representative, and assign $Pr(\phi_i)$ to the representative. If $\min(\phi_i)$ is selected as a representative, then the so computed $F'_{P_{m+1}}(l)$ is greater than $F_{P_{m+1}}(l)$; if $\max(\phi_i)$ is used as a representative, then the so computed $F''_{P_{m+1}}(l)$ is smaller than $F_{P_{m+1}}(l)$.

Example 8.2 (Bucket Approximation). Consider path P_m and edge e_m. Let $f_{P_m|e_m}(x_j|y) = 0.2$ for $1 \leq j \leq 5$. If $t = 2$, then all values in w_{P_m} are divided into three buckets: $[x_1,x_2], [x_3,x_4], [x_5,x_5]$, with probability 0.4, 0.4 and 0.2, respectively.

If the minimal value of each bucket is used as a representative, then x_1, x_3 and x_5 are selected. As a result, the l-weight constrained region of P_{m+1} is increased such that the shaded area is included. So the calculated $F'_{P_{m+1}}(l)$ is greater than the actual $F_{P_{m+1}}(l)$.

If the maximal value of each bucket is used as a representative, then x_2, x_4 and x_5 are selected. The l-weight constrained region of P_{m+1} is decreased such that the shaded area is excluded. So the calculated $F''_{P_{m+1}}(l)$ is smaller than the actual $F_{P_{m+1}}(l)$. ■

Therefore, the average value of $F'_{P_{m+1}}(l)$ and $F''_{P_{m+1}}(l)$ can be used to approximate the actual $F_{P_{m+1}}(l)$. The approximation quality is guaranteed by the following lemma.

Lemma 8.3 (Approximation quality). *Given a real value $l > 0$ and an integer $t > 0$, let $\{\phi_i = [x_{z_i}, x_{z'_i}]\}$ be a set of buckets of w_{P_m}. Let $F'_{P_{m+1}}(l)$ and $F''_{P_{m+1}}(l)$ be the l-weight probabilities computed using $\{\min(\phi_i)\}$ and $\{\max(\phi_i)\}$, respectively, then,*

$$F'_{P_{m+1}}(l) - F''_{P_{m+1}}(l) \leq \frac{1}{t} \tag{8.4}$$

Moreover,

$$|\widehat{F}_{P_{m+1}}(l) - F_{P_{m+1}}(l)| \leq \frac{1}{2t} \tag{8.5}$$

where $\widehat{F}_{P_{m+1}}(l) = \frac{F'_{P_{m+1}}(l)+F''_{P_{m+1}}(l)}{2}$.

Proof. $F'_{P_{m+1}}(l) - F''_{P_{m+1}}(l)$ is the sum of the probability of the shaded area in Figures 8.3(a) and 8.3(b). In each bucket b_i, the width of the shaded area is $Pr(b_i) < \frac{1}{t}$. The length sum of all pieces of shaded are is at most 1. Thus, the probability of the shaded area is at most $\frac{1}{t} \times 1 = \frac{1}{t}$. Inequality 8.4 is proved.

Conclusion 8.5 is derived from Inequalities 8.4 directly. ∎

After obtaining the approximate probability distribution of $w_{P_{m+1}}$, we can further divide the approximate $w_{P_{m+1}}$ into buckets, and compute the approximate probability distribution of $w_{P_{m+2}}$. By applying the bucket approximation iteratively, we finally obtain an approximate l-weight probability of path P, and the approximation quality is guaranteed by the following theorem.

Theorem 8.2 (Overall approximation quality). *Given a real value $l > 0$, an integer $t > 0$, and a path $P = \langle v_1, \ldots, v_{m+1} \rangle$ containing m edges, let $F_P(l)$ be the exact l-weight probability and $\widehat{F}_P(l)$ be the approximate l-weight probability computed using iterative bucket approximation, then $|\widehat{F}_P(l) - F_P(l)| \leq \frac{m-1}{2t}$.*

Proof. P contains m edges, so $m-1$ steps of bucket approximation are needed to compute the probability distribution of P. We prove the theorem using Mathematical Induction.

In the first step (computing the probability of $P_3 = \langle v_1, v_2, v_3 \rangle$), we have $|\widehat{F}_{P_3}(l) - F_{P_3}(l)| \leq \frac{1}{2t}$, which is shown in Lemma 8.3.

Suppose the conclusion holds for the $(j-1)$-th step (computing the probability of P_{j+1}). That is,

$$|\widehat{F}_{P_{j+1}}(l) - F_{P_{j+1}}(l)| \leq \frac{j-1}{2t} \tag{8.6}$$

for any real value $l > 0$.

To compute the probability distribution of $w_{P_{j+2}}$, the approximate weight of $w_{P_{j+1}}$ is divided into buckets, such that the probability of each bucket $b_i = [x_i, x'_i]$ is at most $\frac{1}{t}$. Since the buckets are constructed based on the approximation probability distribution of P_{j+1}, we have $\widehat{Pr}(b_i) = \widehat{F}_{P_{j+1}}(x'_i) - \widehat{F}_{P_{j+1}}(x'_{i-1}) \leq \frac{1}{t}$. From Inequality 8.6, we have $|\widehat{F}_{P_{j+1}}(x'_i) - F_{P_{j+1}}(x'_i)| \leq \frac{j-1}{2t}$ and $|\widehat{F}_{P_{j+1}}(x'_{i-1}) - F_{P_{j+1}}(x'_{i-1})| \leq \frac{j-1}{2t}$, the actual probability of b_i is $Pr(b_i) = F_{P_{j+1}}(x'_i) - F_{P_{j+1}}(x'_{i-1}) = \frac{1}{t} + \frac{j-1}{2t} \times 2 = \frac{j}{t}$. Using the similar proof of Lemma 8.3, the approximation quality of P_{j+2} (the j-th step) can be derived as $|\widehat{F}_{P_{j+2}}(x) - F_{P_{j+2}}(x)| \leq \frac{j}{2t}$.

To compute the distribution of P, there are overall $m-1$ steps. Thus, the conclusion holds. ■

The complexity is analyzed as follows. Lemma 8.2 shows that there are at most $2t$ buckets constructed in the approximation. Therefore, calculating the approximate $f_{P_{i+1}}(x)$ from f_{P_i} and $f_{e_i}(x)$ $(i \geq 2)$ takes $O(t \times |w_{e_i}|)$ time. The overall complexity of computing $f_{m+1}(x)$ is $O(t \sum_{2 \leq i \leq m} |w_{e_i}|)$.

8.1.3 Estimating *l*-Weight Probabilities

The l-weight probability of a path can also be estimated using sampling. For a path $P = \langle v_1, \ldots, v_{m+1} \rangle$, let X_P be a random variable as an indicator to the event that $w_P \leq l$. $X_P = 1$ if $w_P \leq l$; $X_P = 0$ otherwise. Then, the expectation of X_P is $E[X_P] = F_P(l)$.

To estimate $E[X_P]$, we draw a set of samples uniformly with replacement. Each sample unit s is an observation of the path weight, which is generated as follows. At first, s is set to 0. Then, for edge $e_1 \in P$, we choose a value x_1 in w_{e_1} following the probability distribution $f_{e_1}(x)$. Then, for each edge $e_i \in P$ $(2 \leq i \leq m)$, we choose a value x_i in w_{e_i} following the probability distribution of $f_{e_i|e_{i-1}}(x|x_{i-1})$. The chosen value is added to s. Once the weight values of all edges have been chosen, we compare s with l. If $s \leq l$, then the indicator X_P for s is set to 1, otherwise, it is set to 0.

We repeat the above procedure until a set of samples S are obtained. The mean of X_P in S is $E_S[X_P]$, which can be used to estimate $E[X_P]$. If the sample size is sufficiently large, the approximation quality is guaranteed following with the well known Chernoff-Hoeffding bound [182].

Theorem 8.3 (Sample size). *Given a path P, for any δ $(0 < \delta < 1)$, ε $(\varepsilon > 0)$, and a set of samples S of P, if $|S| \geq \frac{3 \ln \frac{2}{\delta}}{\varepsilon^2}$ then* $\Pr\{|E_S[X_P] - E[X_P]| > \varepsilon\} \leq \delta$.

Proof. The theorem is an immediate application of the well known Chernoff-Hoeffding bound [182]. ■

The complexity of estimating $E[X_P]$ is $O(|S| \cdot |P|)$, where $|S|$ is the number of samples drawn and $|P|$ is the number of edges in P.

8.1.4 A Depth First Search Method

Straightforwardly, the **depth-first path search method** can be used to answer a path query $Q_l^{\tau}(u, v)$. The search starts at u. Each time when a new vertex v_i is visited, the weight probability mass function of the path $P = \langle u, \ldots, v_i \rangle$ is calculated, using one of the three methods discussed in this section. If $v_i = v$ and $F_P(l) \leq \tau$, then P is added to the answer set.

Since weights are positive, as more edges are added into a path, the l-weight probability of the path decreases. Therefore, during the search, if the current path does not satisfy the query, then all its super paths should not be searched, as they cannot be in the answer set.

Lemma 8.4 (Monotonicity). *Given a path P and its subpath P', and a weight threshold $l > 0$, $F_P(l) \leq F_{P'}(l)$.* ■

The overall complexity of the depth-first path search algorithm is $O(\sum_{P \in \mathscr{P}} C(P))$, where $\mathscr{P}$ is the set of visited paths and $C(P)$ is the cost of the l-weight probability calculation. If the exact method is used, then $C(P) = \prod_{e \in P} |w_e|$. If the bucket approximation method is used, then $C(P) = 4t \times \sum_{e \in P} |w_e|$, where t is the bucket parameter. If the sampling method is used, then $C(P) = |S| \times |P|$, where $|S|$ is the number of samples and $|P|$ is the number of edges in P.

The method can be extended to answer top-k path queries as following. To answer a WT top-k path query, the probability threshold τ is set to 0 at the beginning. Once a path between u and v is found, we compute its l-weight probability, and add the path to a buffer. If the buffer contains k paths, we set τ to the smallest l-weight probability of the paths in the buffer. During the search, when a new path is found between u and v and satisfying the threshold τ, it is added into the buffer. At the same time, the path in the buffer with the smallest l-weight probability is removed from the buffer. τ is updated accordingly. Therefore, during the search, the buffer always contains at most k paths. At the end of the search, the k paths in the buffer are returned.

A PT top-k path query can be answered following the similar procedure. We set the weight threshold $l = +\infty$ at the beginning. During the search, a buffer always stores at most k found paths between u and v with the smallest τ-confident weights. l is set to the value of the smallest τ-confident weight of the paths in the buffer. The k paths in the buffer at the end of the search are returned as the answers.

Limited by space, hereafter, we focus on answering probabilistic path queries and omit the details for top-k path query evaluation.

8.2 P*: A Best First Search Method

Although the approximation algorithms presented in Section 8.1 to accelerate the probability calculation, it is still computationally challenging to search for all paths in real road networks. In this section, we present the **P* algorithm**, a best first search algorithm for efficient probabilistic path query evaluation

P* carries the similar spirit as the A* algorithm [150, 151]. It visits the vertex that is most likely to be an answer path using a heuristic evaluation function, and stops when the rest unexplored paths have no potential to satisfy the query. However, the two algorithms are critically different due to the different types of graphs and queries.

A* is used to find the shortest path between two vertices u and v in a certain graph. Therefore, the heuristic evaluation function for each vertex v_i is simply the sum of the actual distance between u and v_i and the estimated distance between v_i and v.

P* aims to find the paths that satisfy the weight threshold l and probability threshold p between two vertices u and v in a probabilistic graph with complex correlations among edge weights. Therefore, the heuristic evaluation function for each vertex v_i is the complex joint distribution on a set of correlated random variables. This posts serious challenges in designing heuristic evaluation functions and calculation.

In this section, we first introduce the intuition of the P* algorithm. Then, we design three heuristic evaluation functions that can be used in the P* algorithm. In Section 8.3, a hierarchical index is developed to support efficient query answering using the P* algorithm.

8.2.1 The P Algorithm*

To answer a probabilistic path query $Q_l^\tau(u,v)$, we search the paths from vertex u. The situation during the search is illustrated in Figure 8.4. Generally, before we decide to visit a vertex v_i, we want to evaluate how likely v_i would be included in an answer path. Suppose P_1 is the explored path from u to v_i, and P_2 is an unexplored path from v_i to v, then the probability distribution of the path P from u to v that contains P_1 and P_2 is given by the following theorem.

Theorem 8.4 (Super path probability). *Let $P_1 = \langle u,\ldots,v_i\rangle$, $P_2 = \langle v_i,\ldots,v\rangle$ such that $(P_1 \cap P_2 = \{v_i\})$ and $e = (v_{i-1},v_i)$, the l-weight probability of super path $P = \langle u,\ldots,v_i,\ldots,v\rangle$ is*

$$F_P(l) = \sum_{x_1+y+x_2 \le l} f_{P_1,e}(x_1,y) \times f_{P_2|e}(x_2|y) \tag{8.7}$$

Proof. The weight of P is $w_P = w_{P_1} + w_{P_2}$. Therefore,

$$\begin{aligned} F_P(l) &= Pr[w_P \le l] = Pr[w_{P_1} + w_{P_2} \le l] \\ &= \textstyle\sum_{x_1+x_2 \le l} Pr[w_{P_1} = x_1, w_{P_2} = x_2] \\ &= \textstyle\sum_{x_1+x_2 \le l} Pr[w_{P_1} = x_1] Pr[w_{P_2} = x_2 | w_{P_1} = x_1] \end{aligned}$$

Since w_{P_1} and w_{P_2} are conditionally independent given w_e, we have

$$Pr[w_{P_2} = x_2 | w_{P_1} = x_1] = \textstyle\sum_{y \le x_1} Pr[w_{P_2} = x_2 | w_e = y]$$

Equation 8.7 follows directly. ∎

In Equation 8.7, $f_{P_1,e}(x_1,y)$ can be easily calculated according to Theorem 8.1. It also can be computed using the approximation methods discussed in Section 8.1. However, $f_{P_2|e}(x_2|y)$, the probability distribution of P_2 given edge e, is unknown.

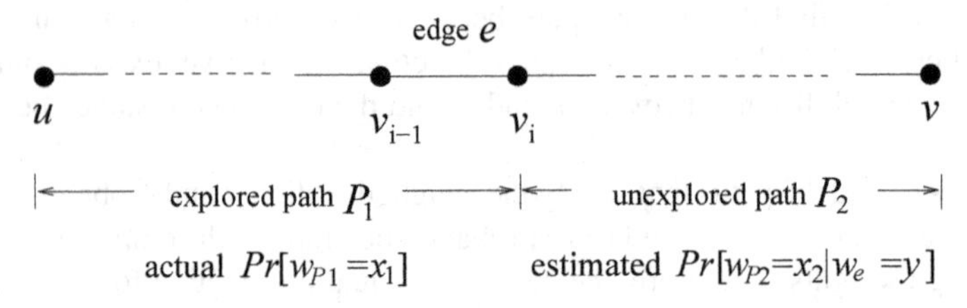

Fig. 8.4 P* search process.

The objective of P* is to find a good **heuristic estimate** $h_{P_2|e}(x_2|y)$ for $f_{P_2|e}(x_2|y)$, such that $F_P(l)$ can be estimated before visiting v_i. Then, the vertex with the higher estimated $F_P(l)$ is visited earlier. The estimated $F_P(l)$ is used as a **heuristic evaluation function** of vertex v_i:

$$\Delta(v_i,l) = \hat{F}_P(l) = \sum_{x_1+y+x_2 \leq l} f_{P_1,e}(x_1,y) \times h_{P_2|e}(x_2|y) \tag{8.8}$$

In order to answer a query $Q_l^\tau(u,v)$, P* starts from u. For any vertex v_i adjacent to u, P* calculates $\Delta(v_i,l)$ and puts v_i into a priority queue. Each time, P* removes the vertex v_i with the highest $\Delta(v_i,l)$ from the priority queue. After v_i is visited, the Δ scores of other vertices are updated accordingly. A path P is output if v is reached and $F_P(l) \geq \tau$. The algorithm stops if the priority queue is empty, or the Δ scores of the vertices in the priority queue are all smaller than τ.

In order to guarantee that P* finds all answer paths, the heuristic evaluation function should satisfy the following requirement.

Theorem 8.5 (Heuristic evaluation function). *P* outputs all answer paths if and only if for any path* $P = \langle u, \ldots, v_i, \ldots, v \rangle$, $\Delta(v_i,l) \geq F_P(l)$.

Proof. If there is a path P such that $\Delta(v_i,l) < \tau \leq F_P(l)$. Then, P will not be returned by P* but it is actually an answer path. ■

$\Delta(v_i,l)$ is called a **valid heuristic evaluation function** if $\Delta(v_i,l) \geq F_P(l)$.

8.2.2 Heuristic Estimates

It is important to find a good heuristic estimate $h_{P_2|e}(x_2|y)$ so that the evaluation function $\Delta(v_i,l)$ is valid and close to the real $F_P(l)$. In this subsection, we first present two simple heuristic estimates that satisfy Theorem 8.5. Then, we derive a sufficient condition for valid heuristic estimates, and present a more sophisticated heuristic estimate.

8.2.2.1 Constant Estimates

Trivially, we can set

$$h_{P_2|e}(x_2|y) = \begin{cases} 1, & x_2 \geq 0; \\ 0, & \text{otherwise.} \end{cases}$$

Then, the evaluation function becomes

$$\Delta(v_i, l) = \sum_{x_1+y\leq l} f_{P_1,e}(x_1, y) = F_{P_1}(l)$$

In this case, at each step, P* always visits the vertices v_i that maximize the l-weight probability of $P_1 = \langle u, \ldots, v_i \rangle$. The subpaths $\langle u, \ldots, v_j \rangle$ whose current l-weight probability is smaller than τ are pruned.

Example 8.3 (Constant estimates). Consider the probabilistic graph in Figure 2.6. To answer a query $Q_{15}^{0.3}(A,E)$. The search starts from A. Using the constant estimates, *the evaluation functions for B and C are:*

$$\Delta(B, 15) = \sum_{x_1\leq 15} f_{AB}(x_1) = 0.6, \text{ and } \Delta(C, 15) = \sum_{x_1\leq 15} f_{AC}(x_1) = 1.0.$$

Since $\Delta(C, 15)$ is larger, C should be explored first. ■

The constant value estimates are easy to construct. But clearly, the constant value estimates do not consider the relationship between the current vertex to the unvisited end vertex v.

8.2.2.2 Min-Value Estimates

To incorporate more information about the unexplored paths into decision making, we consider the minimal possible weights from the current vertex v_i to the end vertex v. We construct a certain graph G' with the same set of vertices and edges as the uncertain graph G. For each edge e, the weight of e is the minimal value in w_e of G. Let l_i^{min} be the weight of the shortest path between v_i and v, excluding the visited edges. Let the estimation function be

$$h_{P_2|e}(x_2|y) = \begin{cases} 1, & x_2 = l_i^{min}; \\ 0, & \text{otherwise.} \end{cases}$$

The evaluation function becomes

$$\begin{aligned} \Delta(v_i, l) &= \Sigma_{x_1+y+l_i^{min}\leq l} f_{P_1,e}(x_1, y) \times h_{P_2|e}(l_i^{min}|y) \\ &= \Sigma_{x_1+y\leq l-l_i^{min}} f_{P_1,e}(x_1, y) \\ &= F_{P_1}(l - l_i^{min}) \end{aligned}$$

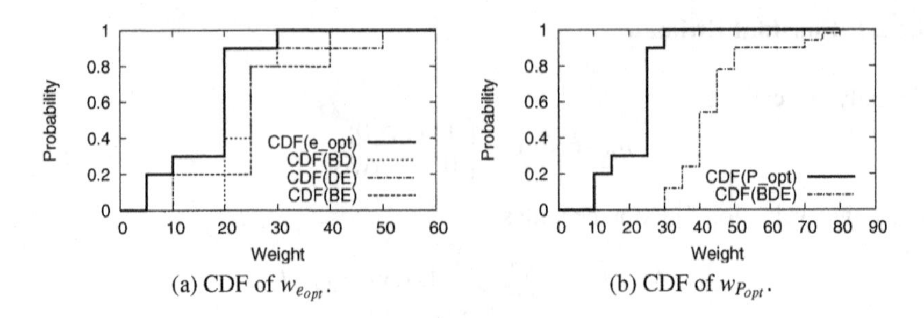

(a) CDF of $w_{e_{opt}}$. (b) CDF of $w_{P_{opt}}$.

Fig. 8.5 CDFs of "virtual optimal" edges and paths.

The search algorithm always visits the vertex v_i that maximizes the $(l - l_i^{min})$-weight probability of $P_1 = \langle u, \ldots, v_i \rangle$. The subpaths $\langle u, \ldots, v_j \rangle$ whose $(l - l_j^{min})$-weight probabilities are smaller than τ are pruned.

Example 8.4 (Min-value estimates). Consider the probabilistic graph in Figure 2.6 and query $Q_{15}^{0.3}(A,E)$ again. Using the min-value estimates, *we construct a certain graph G' with weights $w_{AB} = 5$, $w_{AC} = 5$, $w_{BD} = 20$, $w_{BE} = 5$, $w_{CE} = 10$, and $w_{DE} = 10$.*

The shortest distance from B to E through the unvisited edges in G' is 5. The evaluation function for B is $\Delta(B,15) = \sum_{x_1 \leq 15-5} f_{AB}(x_1) = 0.3$. The shortest distance from C to E in G' is 10. The evaluation function for C is $\Delta(C,15) = \sum_{x_1 \leq 15-10} f_{AC}(x_1) = 0.2$. Since $\Delta(B,15)$ is larger, B should be explored first. Moreover, there is no need to visit C further because $\Delta(C,15) < \tau$. ∎

Compared to the *constant estimates*, the *min-value estimates* consider more information about the unexplored paths, and give priority to the vertices that are closer to the end vertex v and are more likely to satisfy the query. The drawback of the min-value estimates is that they do not consider the probabilistic distribution of unexplored paths.

8.2.2.3 Stochastic Estimates

How can we incorporate the probability distribution information of the unexplored paths in heuristic estimates? Let $\mathscr{P}_2 = \{P_{2_1}, \ldots, P_{2_m}\}$ be all paths from v_i to v that do not contain any visited vertices except for v_i. We can construct a "virtual path" $P_{opt} = \langle v_i, \ldots, v \rangle$ such that for any real number x and $P_{2_i} \in \mathscr{P}_2$, $F_{P_{opt}}(x) \geq F_{P_{2_i}}(x)$.

Definition 8.1 (Stochastic order [191]). For two random variables r_1 and r_2 with distribution functions $F_{r_1}(x)$ and $F_{r_2}(x)$, respectively, r_1 is smaller than r_2 in stochastic order if for any real number x, $F_{r_1}(x) \geq F_{r_2}(x)$. ∎

Definition 8.2 (Stochastic dominance). For two paths P_1 and P_2, P_1 **stochastically dominates** P_2, if w_{P_1}, the weight or P_1, is smaller than w_{P_2}, the weight of P_2, in stochastic order. ■

P_{opt} **stochastically dominates** all paths in $\mathscr{P}_2$, if $w_{P_{opt}}$ is smaller than any w_{P_i} $(1 \leq i \leq m)$ in stochastic order. An example is shown in Figure 8.5(b). P_{opt} stochastically dominates $\langle B, D, E \rangle$. The following theorem shows that P_{opt} can be used to define a valid heuristic evaluation function.

Theorem 8.6 (Sufficient condition). *Let $\mathscr{P}_2 = \{P_{2_1}, \ldots, P_{2_m}\}$ be all paths between v_i and v in graph G. For a path $P_{opt} = \langle v_i, \ldots, v \rangle$, if P_{opt} stochastically dominates all paths in $\mathscr{P}_2$, then*

$$\Delta(v_i, l) = \sum_{x_1+y+x_2 \leq l} f_{P_1,e}(x_1, y) \times f_{P_{opt}}(x_2)$$

is a valid heuristic evaluation function for P.*

Proof. Compare two paths $P = P_1 + P_{opt}$ and $P' = P_1 + P_{2_i}$ $(P_{2_i} \in \mathscr{P}_2)$.

$$Pr[w_P \leq l] = \sum_{x \leq l} Pr[w_{P_1} = x] \times Pr[w_{P_{opt}} \leq l - x]$$

$$Pr[w'_P \leq l] = \sum_{x \leq l} Pr[w_{P_1} = x] \times Pr[w_{P_{2_i}} \leq l - x]$$

Since $Pr[w_{P_{opt}} \leq l - x] \geq Pr[w_{P_{2_i}} \leq l - x]$, we have $Pr[w_P \leq l] \geq Pr[w'_P \leq l]$. ■

More than one P_{opt} can be constructed. Here we present a simple three-step construction method that ensures the resulting path is a stochastically dominating path for $\mathscr{P}_2$.

Step 1: Constructing the path

We find the path P_{2_i} in $\mathscr{P}_2$ with the least number of edges. Let n be the number of edges in P_{2_i}. We construct $n-1$ virtual vertices $\hat{v}_1, \ldots, \hat{v}_{n-1}$. Let $P_{opt} = \langle v_i, \hat{v}_1, \ldots, \hat{v}_{n-1}, v \rangle$. Thus, P_{opt} has the least number of edges among all edges in $\mathscr{P}_2$.

For example, in the probabilistic graph in Figure 2.6, there are two paths between B and E: $P_1 = \langle B, E \rangle$ and $P_2 = \langle B, D, E \rangle$. Since P_1 only contains one edge, the path P_{opt} should also contain only one edge.

Step 2: Assigning edge weights

Let $\mathscr{E}$ be the set of edges in $\mathscr{P}_2$. We want to construct the weight of an edge e_{opt} in P_{opt} such that e_{opt} stochastically dominates all edges in $\mathscr{E}$.

We construct $w_{e_{opt}}$ as follows. At the beginning, we set $w_{e_{opt}} = \emptyset$. Then, we represent each sample $x \in w_e$ ($e \in \mathscr{E}$) as a pair $(x, F_e(x))$. We sort all samples in the value ascending order. If there are two samples with the same value, we only keep the sample with the larger cumulative probability. In the next step, we scan the samples in the sorted list. If $w_{e_{opt}} = \emptyset$, then we add the current sample $(x, F_e(x))$ into $w_{e_{opt}}$. Otherwise, let $(x', F_{e_{opt}}(x'))$ be the sample with the largest value x' in $w_{e_{opt}}$. We add the current sample $(x, F_e(x))$ into $w_{e_{opt}}$ if $x > x'$ and $F_e(x) > F_{e_{opt}}(x')$. Last, we assign weight $w_{e_{opt}}$ to each edge $e_i \in P_{opt}$ $(1 \le i \le n)$.

Example 8.5 (Assigning edge weights). Consider the probabilistic graph in Figure 2.6. There are two paths between B and E: $P_1 = \langle B, E \rangle$ and $P_2 = \langle B, D, E \rangle$. $\mathscr{E} = \{BD, DE, BE\}$. The probability mass function of the constructed weight $w_{e_{opt}}$ is $\{5(0.2), 10(0.1), 20(0.6), 30(0.4)\}$. $w_{e_{opt}}$ is smaller than w_{BD}, w_{DE} and w_{BE} in stochastic order. The weight cumulative distribution functions of those edges are shown in Figure 8.5(a). ■

Step 3: Assigning the path weight

$w_{P_{opt}} = \sum_{1 \le i \le n} w_{e_i}$. The distribution of $w_{P_{opt}}$ depends on the marginal distribution of w_{e_i} and the correlations among w_{e_i}'s. If the correlations among edges are explicitly represented, then we just construct $w_{P_{opt}}$ according to the correlation function.

In most cases, the correlation functions among weights are not available. Therefore, we want to construct a path weight $w_{P_{opt}}$ that stochastically dominates all possible weights given the same $w_{e_1}, \ldots, w_{e_n}$, as defined in the following rule.

Lemma 8.5 (Upper bound). *For path P_{opt} containing edges $e_1, \ldots, e_n$, let x_i^{min} be the smallest sample of weight w_{e_i}, and $l_{min} = \sum_{1 \le i \le n} x_i^{min}$. Then, $F_{P_{opt}}(l) \le \min_{1 \le i \le n}\{F_{e_i}(l - l_{min} + x_i^{min})\}$.*

Proof. The lemma follows with Lemma 8.1 directly. ■

Therefore, for any value l, we assign

$$F_{P_{opt}}(l) = \min_{1 \le i \le n} \{F_{e_i}(l - l_{min} + x_i^{min})\}.$$

The heuristic evaluation function is

$$\Delta(v_i, l) = \sum_{x_1 + y + x_2 \le l} f_{P_1, e}(x_1, y) \times f_{P_{opt}}(x_2).$$

Example 8.6 (Assigning path weights). Continuing Example 8.5, since P_{opt} only contains one edge, the probability distribution of P_{opt} is the same as that of e_{opt}.

As another example, suppose P_{opt} contains two edges e_1 and e_2 with the same weight $w_{e_{opt}}$. How can we calculate the probability distribution of P_{opt}? The sum of minimal samples of w_{e_1} and w_{e_2} is 10. Then, $F_{P_{opt}}(20)$ is $\min\{F_{e_1}(20-10) = 0.3, F_{e_2}(20-10) = 0.3\} = 0.3$. The cumulative distribution function of P_{opt} is shown in Figure 8.5(b). ■

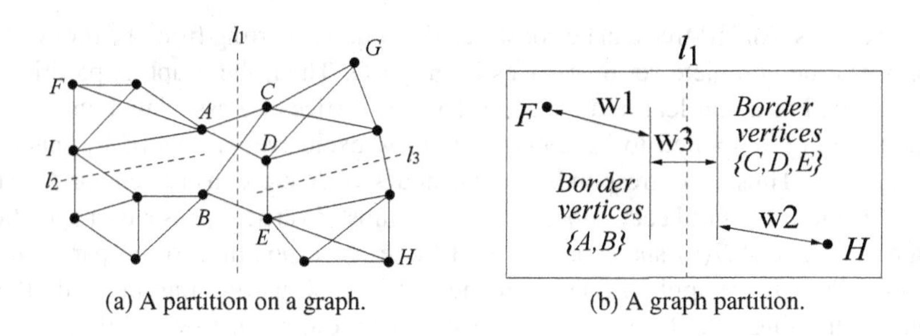

(a) A partition on a graph. (b) A graph partition.

Fig. 8.6 A partition of a graph.

Using the stochastic estimates, in each step, the search algorithm visits the vertex v_i whose heuristic evaluation function using optimal virtual path P_{opt} is the largest. However, constructing P_{opt} requires enumerating all possible paths from v_i to v, which is computationally challenging. Therefore, in the next section, we discuss how to approximate P_{opt} using a hierarchical partition tree index in real road networks.

8.3 A Hierarchical Index for P*

In this section, we introduce a hierarchical partition tree to index the vertices of a graph and maintain the information of weight probability distribution for efficient query answering using the P* algorithm.

8.3.1 HP-Tree Index

A graph partitioning divides the vertices of a graph into subsets of about equal size, such that there are few edges between subsets [192]. Figure 8.6(a) illustrates a 2-partitioning on a graph, where all vertices are divided into two subsets, separated by l_1. Among the vertices on the left of l_1, only A and B are connected to the vertices on the right subset. Therefore, they are called **border vertices**. Similarly, C, D and E are the border vertices in the right subset.

The 2-partitioning can be applied recursively on each subset. As shown in Figure 8.6(a), the left subset is further partitioned into two smaller subsets by line l_2, and the right subset is further partitioned into two smaller subsets by l_3. By recursive partitioning, we can obtain a **hierarchical partition tree** of the graph, as illustrated in Figure 8.7.

Given a graph G and an m-partitioning P on G, a **hierarchical partition tree** (HP-tree for short) is an m-nary tree that indexes the vertices in G according to P. Each leaf node in an HP-tree represents a vertex, and each non-leaf node represents a set of vertices. An HP-tree can be constructed top-down starting from the root which represents the complete set of vertices in graph G. Then, the graph is partitioned recursively. For a non-leaf node N and the set of vertices V_N associated with N, an m-partitioning is applied to V_N and results in m exclusive subsets of V_N, namely $V_{N_1}, \ldots, V_{N_m}$. Thus, m corresponding child nodes of N are constructed for N. The partition continues until each subset contains at most d vertices (d is a user specified number). Figure 8.7(b) shows a 2-nary HP-tree corresponding to the partition in Figure 8.7(a). If we apply a linear time heuristic graph partitioning method [193], then constructing an HP-tree using m-partitioning requires $O(n \log_m \frac{n}{d})$ time.

8.3.2 Approximating Min-Value Estimates

To approximate a min-value estimate, we store auxiliary information for each node in an HP-tree as follows. For each leaf node N_L representing a vertex v and its parent node N associated with a set of vertices V_N, we compute the weight of the shortest path between v and each border vertex of V_N. The smallest weight is stored in N_L. Then, for node N and each of its ancestor nodes N_A, let V_A be the set of vertices associated with N_A. We compute the shortest paths between each border vertex of V_N and each border vertex of V_A. The smallest weight is stored in N.

For example, in the HP-tree shown in Figure 8.7(b), F is a leaf node, and we store the smallest weight of shortest paths from F to the border vertices of $N7$, which is w_1 in Figure 8.7(a). Then, for node $N7$, we compute the smallest weight of the shortest paths between any border vertex in $N7$ and any border vertex in $N3$. Since $N7$ and $N3$ share a common border vertex (I), the smallest weight is 0. The smallest weight of the shortest paths between $N7$ and $N1$ is w_2 in Figure 8.7(a).

The min-value estimate for a vertex u can be approximated as follows. Let N_u and N_v be the parent node of u and v, respectively. Let N be the lowest common ancestor node of N_u and N_v. Then, the weight of any path between u and v is at least $w(u,N_u)+w(N_u,N)+w(v,N_v)+w(N_v,N)$, where $w(u,N_u)$ and $w(v,N_v)$ are the smallest weight from u and v to N_u and N_v, respectively, and $w(N_u,N)$ and $w(N_v,N)$ are the smallest weight from N_u and N_v to N, respectively. Searching the lowest common ancestor node of two vertices takes $O(\log_m \frac{n}{d})$ time.

For example, in Figure 8.7, the lowest common ancestor node of F and G is the root node, which is partitioned by l_1. Thus, the weight of the shortest path between F and G is at least $w_1 + w_2 + w_3 + w_4$. It can be used as the min-value estimate of F with respect to the end vertex G.

8.3.3 Approximating Stochastic Estimates

To approximate a stochastic estimate, we store two pieces of information for each node in the HP-tree.

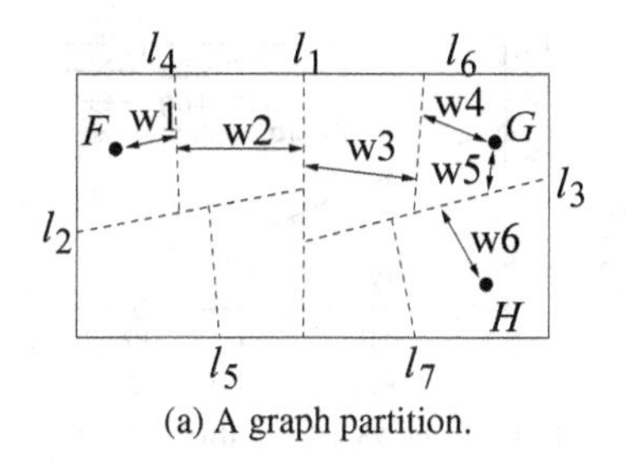

(a) A graph partition.

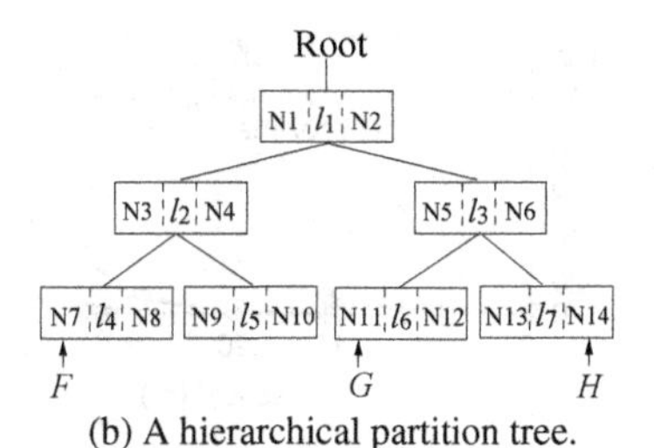

(b) A hierarchical partition tree.

Fig. 8.7 A hierarchical partition of a graph.

For each leaf node N_L representing a vertex v and its parent node N with associated set of vertices V_N, we first compute the shortest edge path between v and each border vertex of V_N. The number of edges is stored in leaf node N_L. Then, we compute a weight that stochastically dominates all weights in V_N, and store it as the "optimal weight" of the node.

For an intermediate node N and each of its ancestor nodes N_A, let V_A be the set of vertices associated with N_A. We first compute the shortest edge paths between each border vertex of V_N and each border vertex of V_A. The number of edges in the path is stored in N. Second, we compute a weight that stochastically dominates all weights of the edges between the border vertices in V_N and V_A. It is stored as the "optimal weight" of the node.

Therefore, the stochastic estimates can be approximated by slightly changing the three steps in Section 8.2.2.3 as follows. In the first step, in order to find the smallest number of edges from a vertex v_i to v, we compute the lower bound of the least number of edges, as illustrated in Figure 8.7. Second, instead of finding an edge weight that stochastically dominates all edges in $\mathscr{P}_2$, we use the optimal weights stored in nodes. The third step remains the same. In this way, we can compute a path that has a smaller number of edges and weights that dominate all weights in P_{opt}.

8.4 Experimental Results

In this section, we report a systematic empirical study. All experiments were conducted on a PC computer with a 3.0 GHz Pentium 4 CPU, 1.0 GB main memory, and a 160 GB hard disk, running the Microsoft Windows XP Professional Edition operating system, Our algorithms were implemented in Microsoft Visual Studio 2005. The graph partition algorithm in METIS 4.0.1[1] and Dijkstra's Shortest Path Algorithm in the Boost C++ Libraries[2] were used in the implementation of HP-trees.

[1] `http://glaros.dtc.umn.edu/gkhome/views/metis`
[2] `http://www.boost.org/`

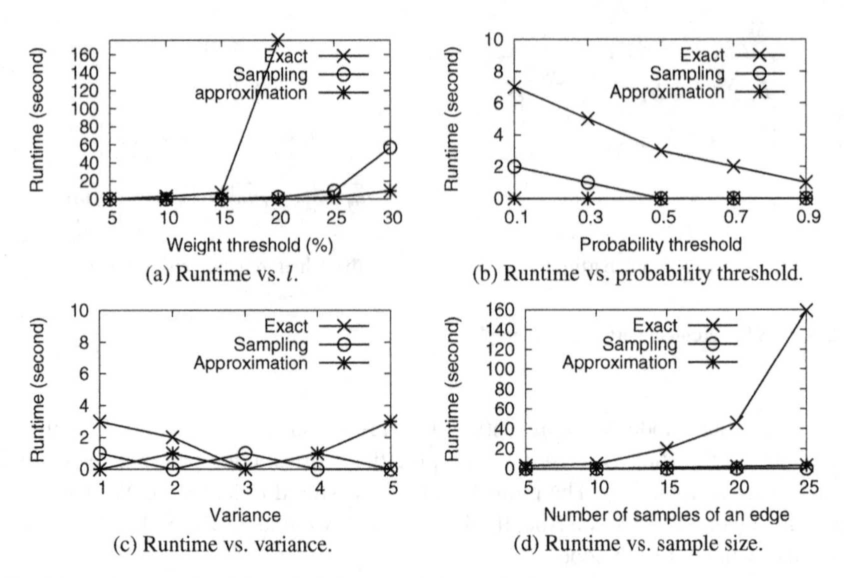

(a) Runtime vs. l.

(b) Runtime vs. probability threshold.

(c) Runtime vs. variance.

(d) Runtime vs. sample size.

Fig. 8.8 Efficiency of weight probability calculation methods.

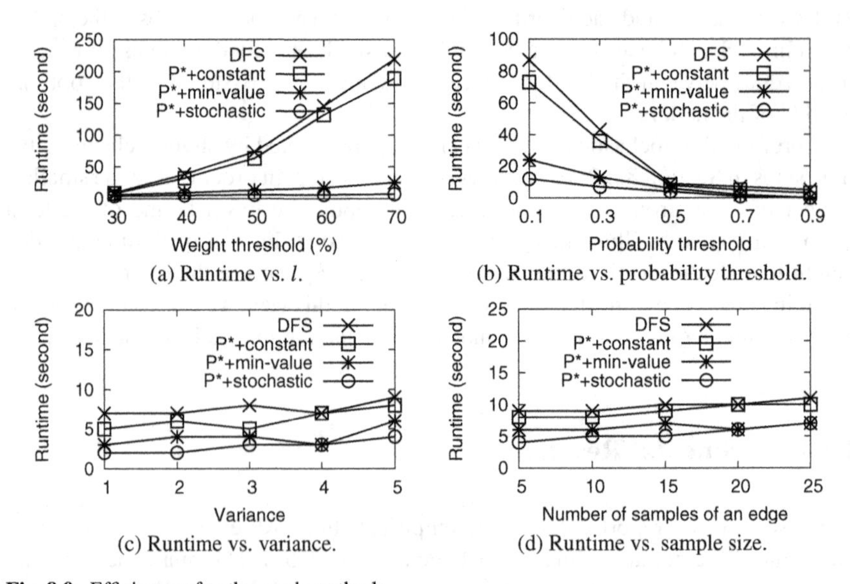

(a) Runtime vs. l.

(b) Runtime vs. probability threshold.

(c) Runtime vs. variance.

(d) Runtime vs. sample size.

Fig. 8.9 Efficiency of path search methods.

8.4.1 Simulation Setup

We test the efficiency, the memory usage, the approximation quality, and the scalability of the algorithms on the following five real road network data sets[3]: City of

[3] http://www.cs.fsu.edu/~lifeifei/SpatialDataset.htm

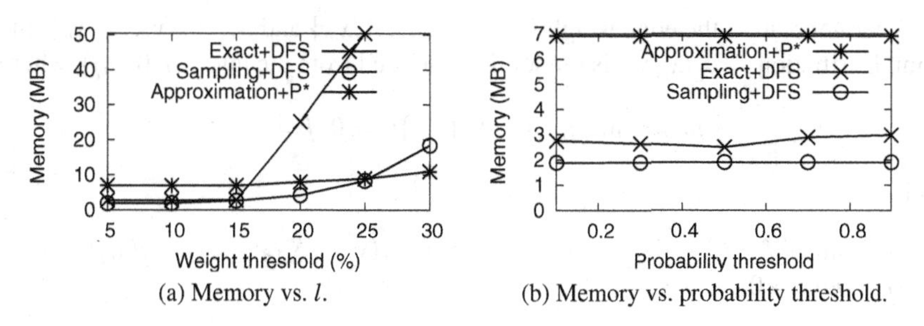

(a) Memory vs. l. (b) Memory vs. probability threshold.

Fig. 8.10 Memory usage.

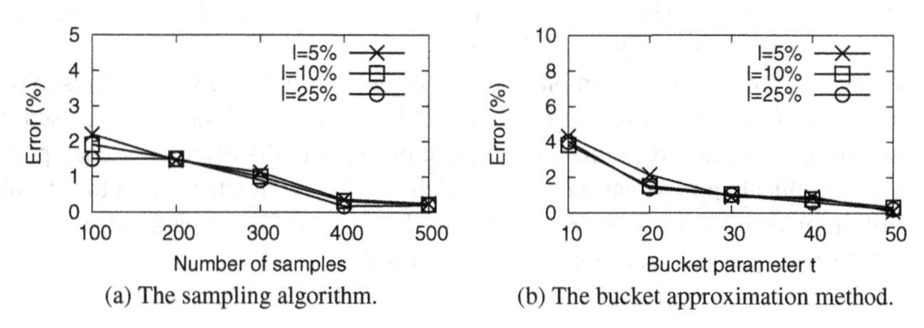

(a) The sampling algorithm. (b) The bucket approximation method.

Fig. 8.11 Approximation quality.

Oldenburg Road Network (*OL*) (6,104 nodes and 7,034 edges), City of San Joaquin County Road Network (*TG*) (18,262 nodes and 23,873 edges), California Road Network (*CAL*) (21,047 nodes and 21,692 edges), San Francisco Road Network (*SF*) (174,955 nodes and 223,000 edges), and North America Road Network (*NA*) (175,812 nodes and 179,178 edges).

The available weight of each edge in the above real road networks is certain. To simulate an uncertain road network, we generate a set of uncertain samples for each edge following the Normal distribution and the Gamma distribution.

In the Normal distribution $N(\mu,\sigma)$, μ is the original edge weight in the certain graphs and σ is the variance. σ is generated for different edges following the Normal distribution $N(\mu_\sigma,\sigma_\sigma)$, where $\mu_\sigma = xR$ ($x = 1\%$ to 5%) and R is the range of weights in the data sets. By default, μ_σ is set to $1\% \cdot R$. This simulation method follows the findings in the existing studies on traffic simulations [194], which indicates that the travel time on paths in road networks follows the Normal distribution in short time intervals.

To simulate the travel time in real road networks using the Gamma distribution $\Gamma(k,\theta)$, as suggested in [195], we can set $\theta = 0.16$ and $k = \frac{\mu}{\theta}$, where μ is the original edge weight in the certain graphs. Since the experimental results on the weights under the Gamma distribution and the Normal distribution are highly similar, limited by space, we only report the results on the data sets with the Normal distribution.

After generating the edge weights $w_{e_1} = \{x_1, \ldots, x_m\}$ and $w_{e_2} = \{y_1, \ldots, y_n\}$, the joint distribution $f_{e_1,e_2}(x_i, y_j)$ is randomly generated from the interval $[0, p_{ij})$, where

$$p_{ij} = \min\{f_{e_1}(x_i), f_{e_2}(y_j)\}(i = 0, j = 0),$$

and

$$p_{ij} = \min\{f_{e_1}(x_i) - \Sigma_{1 \le s \le j-1} f_{e_1,e_2}(x_i, y_s), f_{e_2}(y_j) - \Sigma_{1 \le t \le i-1} f_{e_1,e_2}(x_t, y_j)\}, \\ (i > 0 or j > 0).$$

The path queries are generated as follows. The weight threshold is set to $x\%$ of the diameter d of the network, where d is the maximal weight of the shortest paths among pairs of nodes. The probability threshold varies from 0.1 to 0.9. The start and the end vertices are randomly selected.

By default, the number of samples of each edge is set to 5, the weight threshold l is set to $10\% \cdot d$, and the probability threshold τ is set to 0.5. 500 samples are used in the sampling algorithm to estimate the weight probability distribution of each path. In the hierarchical approximation algorithm, the bucket parameter t is set to 50, and 2-partitionings are used to construct HP-trees. For each different parameter setting, we run 20 path queries and report the average results.

8.4.2 *Efficiency and Memory Usage*

Using data set *OL*, we evaluate the efficiency of the three probability calculation methods, the exact method (*Exact*), the bucket approximation method (*Approximation*), and the sampling based method (*Sampling*). The depth first path search is used. The results are shown in Figure 8.8. Clearly, the bucket approximation method and the sampling based method are more efficient than the exact algorithm.

Particularly, the runtime increases as the weight threshold increases (Figure 8.8(a)), since a larger weight threshold qualifies more paths in the answer set. The runtime of all three algorithms decreases when the probability threshold increases (Figure 8.8(b)), because fewer paths can pass the threshold when the probability threshold is high. The runtime of the algorithms decreases slightly as the variance of the weight samples increases (Figure 8.8(c)). The runtime of the exact algorithm increases significantly when the number of samples of each edge increases, but the runtime of the sampling algorithm and the bucket approximation algorithm remains stable (Figure 8.8(d)) thanks to the efficient approximation probability computation.

We also test the efficiency of the P* search algorithm against the depth first search (*DFS*) (Figure 8.9). Three heuristic estimates are used: constant estimates (*P*+constant*), min-value estimates (*P*+min-value*), and stochastic estimates (*P*+stochastic*). The bucket approximation method is used to compute the path probability distribution. To show the difference in efficiency of the different methods more clearly, we increase the weight threshold to $30\% \cdot d$.

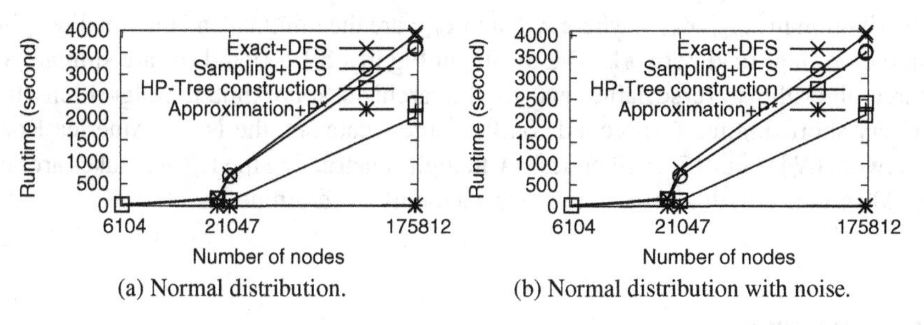

Fig. 8.12 Scalability.

Clearly, the P* search is more efficient than the depth first search method thanks to the heuristic path evaluation during the search. Among the three types of heuristic estimates, stochastic estimates are the most effective, because they use the weight probability distribution of the unexplored paths to guide the search. The HP-tree construction takes 25 seconds on this data set.

The memory usage for different algorithms is shown in Figure 8.10. The memory requirement in the exact probability calculation method with DFS increases rapidly when the weight threshold increases, since longer paths are explored with a larger weight threshold. However, the memory usage in P* search with the bucket approximation probability calculation is stable, because the space used for an HP-tree does not depend on the weight threshold.

8.4.3 Approximation Quality and Scalability

Using data set *OL*, we test the approximation quality of the sampling algorithm and the bucket approximation algorithm. In the same parameter setting as in Figure 8.8, the precision and the recall of all queries are computed. Since they are all 100%, we omit the figures here.

The average approximation error of the l-weight probability computed in the two algorithms is shown in Figure 8.11. For any path P, the approximation error is defined as $\frac{|\hat{F}_P(l)-F_P(l)|}{F_P(l)}$, where $\hat{F}_P(l)$ and $F_P(l)$ are the approximate and exact l-weight probabilities of P, respectively.

The error rate of the sampling method is always lower than 3%, and is very close to 0 when a set of 500 samples are used (Figure 8.11(a)). The average error rate of the bucket approximation method decreases from 4.31% to 0.1% when the bucket parameter t increases from 10 to 50 (Figure 8.11(b)).

Figure 8.12 shows the scalability of the three algorithms on the five real road network data sets. Figure 8.12(a) shows the results for weights following the Normal distribution. Figure 8.12(b) is the results on the data with the Normal distribution and 10% noise. That is, 10% of the edges have a sample drawn from the uniform distribution in $[x_{min}, x_{max}]$, where x_{min} and x_{max} are the minimal and maximal weights in the original road network. The results in Figures 8.12(a) and (b) are similar. All three algorithms are scalable, and the hierarchical approximation algorithms has a very short runtime (20 seconds on the largest data set, the North America Road Network (*NA*) with $175,812$ nodes). Although constructing the HP-tree takes around $2,000$ seconds in this case, it is constructed only once offline.

8.5 Summary

In this chapter, we studied the problem of answering probabilistic path queries defined in Section 2.3.3. There are two major challenges for efficient query evaluation. First, given a path P and a weight threshold l, how can we compute the l-weight probability of P efficiently? Second, given an uncertain road network and a probabilistic path query, how can we search for the paths satisfying the query efficiently?

To solve the first challenge, we developed three methods for efficient l-weight probability calculation: an exact algorithm, a constant factor approximation method and a sampling based approach.

To address the second challenge, we proposed two path search methods.

- A depth-first search algorithm enhanced by effective pruning techniques was developed.
- P* algorithm, a best first search method, was devised to search for all paths satisfying the query. It uses a heuristic method to estimate the l-weight probabilities of the unexplored paths and prune the paths whose estimated l-weight probabilities fail the threshold. Although being an heuristic method, it guarantees to find all paths in the exact solution, since the estimated l-weight probabilities are always no smaller than the actual l-weight probabilities.

An extensive empirical evaluation verified the effectiveness and efficiency of the probabilistic path queries and our methods.

Chapter 9
Conclusions

Uncertain data becomes important and prevalent in many critical applications, such as sensor networks, location-based services, and health-informatics. Due to the importance of those applications and increasing amounts of uncertain data, effective and efficient uncertain data analysis has received more and more attentions from the research community. In this book, we study ranking queries on uncertain data. Particularly, we develop three extended uncertain data models based on the state-of-the-art uncertain data models and propose five novel ranking problems on uncertain data.

In this chapter, we first summarize the book, and then discuss some interesting future directions.

9.1 Summary of the Book

In this book, we study ranking queries on uncertain data and make the following contributions.

- We develop three extended uncertain data models to fit different application interests: the uncertain data stream model, the probabilistic linkage model, and the probabilistic road network model.
- We propose a series of novel *top-k typicality queries* to find the most representative instances within an uncertain object.
  - A *top-k simple typicality* query returns the top-k most typical instances within an uncertain object. The notion of typicality is defined as the likelihood of an instance in the object, which is computed using the kernel density estimation methods.
  - Given two uncertain objects O and S, a *top-k discriminative typicality query* finds the top-k instances in O that are typical in O but atypical in S.
  - A *top-k representative typicality query* finds k instances in an uncertain object that can best represent the distribution of O.

Computing the exact answers to top-k typicality queries is quadratic in nature. We develop three efficient approximation algorithms, a randomized tournament method, a local typicality approximation algorithm that approximates the typicality score of an instance o using a small subset of instances close to o, and a hybrid method that combines the tournament mechanism and the local typicality approximation.

- We develop *probabilistic ranking queries* and *reverse probabilistic ranking queries* on uncertain data by considering the two dimensions: the ranks of uncertain instances or objects and their probabilities of achieving certain ranks.

  - Given a rank parameter k, we can rank instances or objects according to their top-k probabilities (that is, the probabilities of being ranked top-k). A *probabilistic ranking query* finds the uncertain instances or objects whose top-k probabilities are the largest.
  - Given a probability threshold p, the p-rank of an instance o is the minimum k such that $Pr^k(o) \geq p$. Similarly, the p-rank of an object O is the minimum k such that $Pr^k(O) \geq p$. A *reverse probabilistic ranking query* finds the uncertain instances or objects whose p-ranks are the smallest.

  Probabilistic ranking queries and reverse probabilistic ranking queries provide complement views of the ranking characterizes of uncertain data. Moreover, we develop three efficient query evaluation methods, an exact algorithm, a sampling method, and a Poisson approximation based method. Last, we devise an effective and compact index for probabilistic ranking queries and reverse probabilistic ranking queries, which helps efficient online query answering.
- We study the problem of continuous probabilistic top-k query (continuous PT-k query for short) on uncertain streams, which reports, for each time instant t, the set of uncertain streams satisfying the query in the sliding window. This problem extends the ranking queries on static data to dynamic data. To tackle this problem, we develop an exact algorithm that involves several important stream specific techniques and a sampling method. Moreover, we devise the space efficient versions of the exact algorithm and the sampling method that utilize approximate quantiles to maintain the summary of the data streams.
- We study ranking queries on the probabilistic linkage model, which is an extension of the basic uncertain object model by considering the dependencies among objects. Given two sets of tuples A and B, a linkage matches a tuple in A to a tuple in B, meaning that the two tuples refer to the same real world entity. A probability is associated with each linkage, representing the confidence of the match. Each tuple can be associated with multiple linkages, but only one linkage can be true due to the constraint that one tuple can only match at most one tuple in another data set in the ground truth.

  We can consider each tuple $t_A \in A$ as an uncertain object. A tuple $t_B \in B$ can be considered as an instance of t_A if there is a probabilistic linkage $l = (t_A, t_B) \in \mathcal{L}$ such that $Pr(l) > 0$. The membership probability of instance t_B with respect to object t_A is $Pr(l)$. Different from the basic uncertain object model where each instance only belongs to one object, in the probabilistic linkage model, a tuple

$t_B \in B$ may be the instance of multiple objects $\{t_{A_1}, \cdots, t_{A_d}\}$, where t_{A_i} is a tuple in A with linkage $(t_{A_i}, t_B) \in \mathscr{L}$ $(1 \le i \le d)$. A mutual exclusion rule $R_{t_B} = (t_{A_i}, t_B) \oplus \cdots \oplus (t_{A_d}, t_B)$ specifies that t_B should only belong to one object in a possible world. Alternatively, we can consider each tuple $t_B \in B$ as an uncertain object and a tuple $t_A \in A$ is an instance of t_B if there is a linkage $(t_A, t_B) \in \mathscr{L}$.
We develop a probabilistic mutual exclusion graph (PME-graph) to describe the dependencies among objects in the probabilistic linkage model. The PME-graph is shown to be a special form of Markov random fields. Answering ranking queries on probabilistic linkages is significantly different from that on independent uncertain objects. We propose efficient query evaluation methods that are verified to be efficient and scalable by experimental results.

- Last, we extend the ranking uncertain data problem to more complicated data types: uncertain road networks. An uncertain road network is a simple graph, for each edge, the weight is an uncertain object containing a set of instances. As a result, the weight of any path in an uncertain network is also an uncertain object. We propose three types of probabilistic path queries.

  - A *probabilistic path query* with a weight threshold w and a probability threshold p finds the paths between two end vertices whose weights are at most l with a probability of at least p.
  - Given a path P and a weight constraint l, the l-weight probability is the probability that P's weight is at most l. A *weight-threshold top-k path query* returns the paths between two end vertices whose l-weight probabilities are the highest.
  - Given a certain probability threshold p, we can find, for each path, the smallest weight achievable with a probability of at least p. This is called the p-confidence weight of the path. A *probability-threshold top-k path query* finds the paths whose p-confidence weights are the smallest.

  Each type of path queries finds its edge in real applications. To answer those queries efficiently, we address two major challenges. First, we propose three efficient methods to compute l-weight probabilities and p-confidence weights. Then, we develop two path search methods that find the paths satisfying the query efficiently.

9.2 Future Directions: Possible Direct Extensions

It is interesting to extend the ranking queries and their evaluation methods developed in this book to other related uncertain data models and queries. Some of them are listed below.

9.2.1 Top-k typicality queries for uncertain object

In this book, we study three types of top-k typicality queries on static uncertain objects. As future study, it is interesting to extend this study in two directions. First, how can we answer top-k typicality queries on different types of uncertain data, such as uncertain data streams, that is, uncertain objects with evolving distributions? Second, how can we develop alternative of typicality notions that fit different application needs? For example, when the dimensionality of uncertain objects is high, finding the most typical instances in the full space may not be informative. Instead, it is interesting to find in which subspace an instance is typical.

9.2.2 Top-k queries on probabilistic databases

Chapter 5 discusses the problem of top-k query evaluation on probabilistic databases. It is interesting to investigate the following two extensions.

First, the probabilistic database model we adopted in Chapter 5 only considers the generation rules that specify the mutual exclusiveness among tuples. More complex generation rules can be considered. For example, *mutual inclusion rules* that specify the coexistence of the tuples involved in the same rule are discussed in [7]. A *mutual inclusive rule* $R_{\equiv} : t_{r_1} \equiv \cdots \equiv t_{r_m}$ restricts that, among all tuples $t_{r_1}, \cdots, t_{r_m}$ involved in the same rule, either no tuple appears or all tuples appear in a possible world. All tuples in $R_{\equiv}$ have the same membership probability value, which is also the probability of rule $R_{\equiv}$. How to answer top-k queries on probabilistic databases with various generations rules is an interesting extension.

Second, how to incrementally update top-k query results when changes happen to probabilistic databases? For example, the membership probability of an uncertain tuple may be updated when more knowledge about the data is obtained. Instead of recomputing the results based on the updated probabilistic database, it is more efficient to reuse the results computed before updates happen.

9.2.3 Probabilistic ranking queries

In this book, we study probabilistic ranking queries on probabilistic databases, uncertain streams, and probabilistic linkages. It is interesting to extend our methods to other probabilistic data models and other ranking queries. Particularly, there are four important directions.

9.2.3.1 Uncertain data streams with non-uniformly distributed instances

In Chapter 6, we assume that the membership probabilities of all instances in a sliding window are identical. This is suitable for the applications where instances are generated using simple random sampling. However, in other applications, instances with non-identical membership probabilities may be generated. Therefore, it is important to investigate how to adapt the methods developed in Chapter 6 for the case of non-identical membership probabilities.

In the exact algorithm, all techniques can be used except for the "compatible dominant set" technique. The "compatible dominant set" technique reuse the dominant set of an instance in the previous sliding window, as long as the number of instances from each object in the dominant set does not change. This only holds when the membership probabilities of instances are identical. Therefore, new techniques that can reuse the dominant set of instances in the case of non-identical membership probabilities need to be developed.

Second, the sampling method can be used without any change. We simply draw samples according to the distribution of instances for each object in the sliding window.

Last, in order to use the space efficient algorithms developed in Chapter 6, the ϕ-quantile summary needs to be redefined. The major idea is to partition the instances ranked in the value ascending order into intervals, so that the sum of membership probabilities of instances in each interval is at most ϕ. For an uncertain object $W^t_\omega(O)$ containing a set of instances $o_1, \cdots, o_\omega$, where each instance o_i is associated with a membership probability $Pr(o_i)$ $(1 \leq i \leq \omega)$. The instances are ranked in the value ascending order. We partition $o_1, \cdots, o_\omega$ into b exclusive intervals $t_i = [o_{z_i}, o_{z'_i}]$, where

$$\begin{cases} z_i = 1, & i = 1; \\ z_i = z'_{i-1} + 1, & 1 < i \leq b; \\ z'_i = \max\limits_{j \geq z_i} \left\{ j \,\middle|\, \sum\limits_{x=z_i}^{j} Pr(o_x) \leq \phi \right\}, & 1 \leq i \leq b. \end{cases}$$

Therefore, the probability of each interval is at most ϕ. Moreover, there are at most $2\lceil \frac{1}{\phi} \rceil$ intervals. This is because, in the worst case, each interval only contains one instance, which means that the sum of membership probabilities of any two consecutive instances is greater than ϕ. Then, if the number of instances is greater than $2\lceil \frac{1}{\phi} \rceil$, the sum of membership probabilities of all instances will be greater than 1, which conflicts with the fact that the sum of membership probabilities of all instances of one object is 1.

Though the approximate top-k probabilities can be computed similarly as discussed in Chapter 6, details need to be worked out as future work.

9.2.3.2 Uncertain data streams with correlations

Correlations may exist among uncertain data streams. For example, in the speed monitoring application, if two speed sensors are deployed at the same location, then the readings of the two sensors are mutually exclusive. That is, only the reading of one sensor can exist in a possible world. More generally, complex correlation among two or more uncertain data streams can be represented by the joint distribution of their readings. How to continuously monitor the top-k uncertain data streams in such cases is highly interesting.

9.2.3.3 Continuously monitoring probabilistic threshold top-k queries with different parameter values

There are two parameters in probabilistic threshold top-k queries, the query parameter k and the probability threshold p. In some applications, we may be interested in how query results change as parameters vary. To support the interactive analysis, it is highly desirable to monitor probabilistic threshold top-k queries on uncertain data streams with different parameters. To achieve this goal, it is interesting to study if we can extend the *PRist+* index developed in Section 5.5 to uncertain data streams.

9.2.3.4 Continuously monitoring probabilistic threshold top-k aggregate queries

In this chapter, we focus on probabilistic threshold top-k selection queries, where the ranking function is applied on a single instance. Another category of top-k queries is top-k aggregate queries [29], where the ranking function is applied on a group of instances. The top-k groups with highest scores are returned as results. It is interesting to investigate how to extend the techniques discussed in this chapter to handle top-k aggregate queries on uncertain data streams.

9.2.4 Probabilistic path queries on road networks

An important future direction of probabilistic path queries on uncertain road networks is to explore the temporal uncertainty and correlations of travel time along road segments.

The HP-tree can be maintained incrementally as edge weights change. For each node in an HP-tree that contains a set of edges, an optimal weight stochastically dominating all edge weights is stored. The optimal weight can be constructed using a set of (value, probability) pairs that are the skyline points among the (value, probability) pairs of all edges. Therefore, the optimal weight in a sliding window can be maintained using any efficient skyline maintenance algorithms for streams, such as [196].

9.3 Future Directions: Enriching Data Types and Queries

It is highly interesting to study more general types of ranking queries on complex uncertain data. We give some examples here.

9.3.1 Handling Complex Uncertain Data and Data Correlations

In some applications, data tends to be heterogeneous, semi-structured or unstructured. For example, in medical applications, information collected about patients may involve tabular data, text, images, and so on. Uncertainty may widely exist in those applications due to factors like equipment limitations and ambiguity in natural language presentation. Such uncertain data types pose grand challenges in data analytics.

First, it is difficult to model uncertainty in complex data types. For example, documents extracted by hand-writing recognition may be prone to mistakes. How can we represent the uncertainty in each word or prase in documents? Neither the probabilistic database model nor the uncertain object model can be directly adopted.

Second, ranking queries on complex uncertain data may have different forms from the ranking queries on simple uncertain data. In this book, we extend top-k selection queries on uncertain data, where a score is computed for each data record on a set of the attributes. In many applications, complex uncertain data objects may be ranked using different methods. For example, a set of documents or images are often ranked according to their relevance scores to a set of key words. Answering such complicated ranking queries on complex uncertain data is non-trivial.

Last, complex data correlations may exist among uncertain data objects. It is difficult to model the correlations, not to mention answering queries on uncertain data with complex correlations.

9.3.2 Answering More Types of Ranking and Preference Queries on Uncertain Data

In this book, we focus on top-k selection queries on uncertain data, where a total order on all objects or instances is available. More generally, given a set of partial orders on tuples as user preferences, a preference query [197] finds the tuples that best match the preferences. For example, when searching for used cars, a user may specifies his/her preferences as "I like Toyota better than Ford" and "I prefer low mileage". Then we should return all used cars that best match the user's preference. Although there are many studies on preference queries on deterministic data [198, 199, 200, 201, 197], to the best of our knowledge, there are very limited related studies on preference queries on uncertain data.

Let us take skyline queries, an important category of preference queries, as an example. In a deterministic data set, given two tuples t_1 and t_2 in attributes $A_1, \cdots, A_m$, t_1 dominates t_2 if $t_1.A_i \prec t_2.A_i$ $(1 \leq i \leq m)$, where $\prec$ denotes the preference order in attribute a_i. A skyline query returns the tuples that are not dominated by any other tuple in the data set. Pei *et al.* [21] extend skyline queries from deterministic data to the basic uncertain object model. Given a set of independent uncertain objects, the skyline probability of an object O is the probability that O is a skyline object in possible worlds. Given a probability threshold p, a probabilistic threshold skyline query finds the objects whose skyline probabilities are at least p.

As future study, it is interesting to examine how to answer probabilistic threshold skyline queries on the three extended uncertain data models proposed in this work. First, a continuous probabilistic threshold skyline query on a set of uncertain data streams returns, for each time instant, the objects whose skyline probabilities in the sliding window are at least p. This is useful for applications where data is dynamic and uncertain in nature.

Moreover, the probabilistic linkage model proposed in our study is an extension of the basic uncertain object model by considering inter-object dependencies. Skyline queries are useful in some applications of the probabilistic linkage model. For instance, in Example 2.12, a medical expert may be interested in finding the skyline patients in two attributes *age of hospitalization* and *age of death*. It is non-trivial to compute object skyline probabilities in the probabilistic linkage model due to the inter-object dependencies.

Last, in the uncertain road network model, there are multiple paths between two end vertices specified by users. Sometimes, we may want to find the skyline paths that have short travel time and short geographic distances. Therefore, probabilistic skyline queries are highly useful in such applications.

References

1. C. Re, D. Suciu, in *CIKM* (2007), pp. 3–8
2. A.D. Sarma, X. Dong, A. Halevy, in *Managing and Mining Uncertain Data*, ed. by C. Aggarwal (Springer,, 2008), pp. 185–217
3. G. Das, D. Gunopulos, N. Koudas, D. Tsirogiannis, in *VLDB* (2006), pp. 451–462
4. R. Fagin, A. Lotem, M. Naor, in *PODS '01: Proceedings of the twentieth ACM SIGMOD-SIGACT-SIGART symposium on Principles of database systems* (ACM Press, 2001), pp. 102–113. DOI http://doi.acm.org/10.1145/375551.375567
5. K. Mouratidis, S. Bakiras, D. Papadias, in *SIGMOD Conference* (2006), pp. 635–646
6. S. Nepal, M. Ramakrishna, in *Proceedings of the 15th International Conference on Data Engineering* (IEEE Computer Society, 1999), pp. 22–29
7. A.D. Sarma, O. Benjelloun, A. Halevy, J. Widom, in *ICDE '06: Proceedings of the 22nd International Conference on Data Engineering (ICDE'06)* (IEEE Computer Society, Washington, DC, USA, 2006), p. 7. DOI http://dx.doi.org/10.1109/ICDE.2006.174
8. O. Benjelloun, A.D. Sarma, A. Halevy, J. Widom, in *VLDB'2006: Proceedings of the 32nd international conference on Very large data bases* (VLDB Endowment, 2006), pp. 953–964
9. X. Dong, A.Y. Halevy, C. Yu, in *VLDB '07: Proceedings of the 33rd international conference on Very large data bases* (VLDB Endowment, 2007), pp. 687–698
10. P. Agrawal, O. Benjelloun, A.D. Sarma, C. Hayworth, S.U. Nabar, T. Sugihara, J. Widom, in *VLDB* (2006), pp. 1151–1154
11. N.N. Dalvi, D. Suciu, in *Proceedings of the Thirtieth International Conference on Very Large Data Bases* (Toronto, Canada, 2004), pp. 864–875
12. N. Dalvi, D. Suciu, in *Proceedings of the twenty-sixth ACM SIGMOD-SIGACT-SIGART symposium on Principles of database systems (PODS'07)* (ACM, New York, NY, USA, 2007), pp. 1–12. DOI http://doi.acm.org/10.1145/1265530.1265531
13. D. Dey, S. Sarkar, ACM Trans. Database Syst. **21**(3), 339 (1996). DOI http://doi.acm.org/10.1145/232753.232796
14. D. Barbará, H. Garcia-Molina, D. Porter, IEEE Trans. Knowl. Data Eng. **4**(5), 487 (1992)
15. R. Cavallo, M. Pittarelli, in *VLDB* (1987), pp. 71–81
16. T.J. Green, V. Tannen, IEEE Data Eng. Bull. **29**(1), 17 (2006)
17. M.A. Soliman, I.F. Ilyas, K.C.C. Chang, in *Proceedings of the 23nd International Conference on Data Engineering (ICDE'07)* (IEEE, Istanbul, Turkey, 2007)
18. R. Cheng, D.V. Kalashnikov, S. Prabhakar, in *Proceedings of the 2003 ACM SIGMOD international conference on Management of data (SIGMOD'03)* (ACM Press, New York, NY, USA, 2003), pp. 551–562. DOI http://doi.acm.org/10.1145/872757.872823
19. R. Cheng, Y. Xia, S. Prabhakar, R. Shah, J.S. Vitter, in *VLDB* (2004), pp. 876–887
20. Y. Tao, R. Cheng, X. Xiao, W.K. Ngai, B. Kao, S. Prabhakar, in *VLDB* (2005), pp. 922–933
21. J. Pei, B. Jiang, X. Lin, Y. Yuan, in *Proceedings of the 33rd International Conference on Very Large Data Bases (VLDB'07)* (Viena, Austria, 2007)
22. K. Goebel, A.M. Agogino, Journal of Dynamic Systems, Measurement, and Control **123**(1), 145 (2001)
23. S. Abiteboul, P. Kanellakis, G. Grahne, in *Proceedings of the 1987 ACM SIGMOD international conference on Management of data (SIGMOD'87)* (ACM Press, New York, NY, USA, 1987), pp. 34–48. DOI http://doi.acm.org/10.1145/38713.38724
24. T. Imielinski, J. Witold Lipski, Journal of ACM **31**(4), 761 (1984). DOI http://doi.acm.org/10.1145/1634.1886
25. D. Gardner, R. Bostrom. Ohio and kentucky approach to data archiving in cincinnati. URL `http://trb.mtc.ca.gov/urban/adus/Ohio%20and%20Kentucky%20ADUS.pdf`
26. P.M. Lee, *Bayesian statistics: an introduction* (Edward Arnold Publishers Ltd, 1997)
27. A.W.F. Edwards, *Likelihood* (Cambridge University Press; 1 edition, 1985)
28. M. Charikar, S. Guha, E. Tardos, D.B. Shmoys, in *Proceedings of the Symposium on Theory of Computing* (ACM Press, 1999), pp. 1 – 10

29. I.F. Ilyas, G. Beskales, M.A. Soliman, ACM Comput. Surv. **40**(4), 1 (2008). DOI http://doi.acm.org/10.1145/1391729.1391730
30. M. Hua, J. Pei, W. Zhang, X. Lin, in *Proc. International Conference on Data Engineering (ICDE'08)* (Cancun, Mexico, 2008)
31. M. Hua, J. Pei, W. Zhang, X. Lin, in *Proc. ACM International Conference on Management of Data (SIGMOD'08)* (Vancouver, Canada, 2008)
32. A.S. Silberstein, R. Braynard, C. Ellis, K. Munagala, J. Yang, in *ICDE '06: Proceedings of the 22nd International Conference on Data Engineering* (IEEE Computer Society, Washington, DC, USA, 2006), p. 68. DOI http://dx.doi.org/10.1109/ICDE.2006.10
33. X. Zhang, J. Chomicki, in *Proc. the Second International Workshop on Ranking in Databases (DBRank'08)* (2008)
34. G. Kalton, *Introduction to Survey Sampling*, 1st edn. (Sage Publications, Inc., 1983)
35. A. Doucet, B.N. Vo, C. Andrieu, M. Davy, in *Proceedings of the Fifth International Conference on Information Fusion* (2002), pp. 474 – 481
36. B. Svartbo, L. Bygren, T. Gunnarsson, L. Steen, M. Ribe, International Journal of Health Care Quality Assurance **12**, 13 (1 January 1999). URL `http://www.ingentaconnect.com/content/mcb/062/1999/00000012/00000001/art00002`
37. N. Koudas, S. Sarawagi, D. Srivastava, in *SIGMOD '06: Proceedings of the 2006 ACM SIGMOD international conference on Management of data* (ACM, New York, NY, USA, 2006), pp. 802–803. DOI http://doi.acm.org.proxy.lib.sfu.ca/10.1145/1142473.1142599
38. L. Gu, R. Baxter, D. Vickers, C. Rainsford, Record linkage: Current practice and future directions. Tech. Rep. 03/83, CSIRO, the Commonwealth Scientific and Industrial Research Organisation, Australia (2003)
39. L. Antova, C. Koch, D. Olteanu, in *Proceedings of the 2007 IEEE 23rd International Conference on Data Engineering (ICDE'07)* (2007)
40. S.K. Lee, in *Proceedings of the Eighth International Conference on Data Engineering (ICDE'92)* (IEEE Computer Society, Washington, DC, USA, 1992), pp. 614–621
41. P. Sen, A. Deshpande, in *Proceedings of the 23rd IEEE International Conference on Data Engineering (ICDE'07)* (IEEE Computer Society, Washington, DC, USA, 2007), pp. 836–845
42. A. Deshpande, L. Getoor, P. Sen, in *Managing and Mining Uncertain Data*, ed. by C. Aggarwal (Springer,, 2008), pp. 77–105
43. B. Kanagal, A. Deshpande, in *ICDE* (2009)
44. B. Kanagal, A. Deshpande, in *SIGMOD* (2009). URL `http://www.cs.umd.edu/\~{}amol/papers/sigmod09.pdf`
45. P. Sen, A. Deshpande, L. Getoor, Proc. VLDB Endow. **1**(1), 809 (2008). DOI http://doi.acm.org/10.1145/1453856.1453944
46. P. Sen, A. Deshpande, L. Getoor, VLDB Journal, special issue on uncertain and probabilistic databases (2009)
47. S. Singh, C. Mayfield, S. Mittal, S. Prabhakar, S. Hambrusch, R. Shah, in *SIGMOD '08: Proceedings of the 2008 ACM SIGMOD international conference on Management of data* (ACM, New York, NY, USA, 2008), pp. 1239–1242. DOI http://doi.acm.org/10.1145/1376616.1376744
48. S. Singh, C. Mayfield, R. Shah, S. Prabhakar, S.E. Hambrusch, J. Neville, R. Cheng, in *ICDE* (2008), pp. 1053–1061
49. N. Dalvi, D. Suciu, in *Proceedings of the twenty-sixth ACM SIGMOD-SIGACT-SIGART symposium on Principles of database systems (PODS'07)* (ACM Press, New York, NY, USA, 2007), pp. 293–302. DOI http://doi.acm.org/10.1145/1265530.1265571
50. A.L.P. Chen, J.S. Chiu, F.S.C. Tseng, IEEE Trans. on Knowl. and Data Eng. **8**(2), 273 (1996). DOI http://dx.doi.org/10.1109/69.494166
51. C. Ré, N. Dalvi, D. Suciu, in *Proceedings of the 23nd International Conference on Data Engineering (ICDE'07)* (IEEE, Istanbul, Turkey, 2007)

52. D. Burdick, P.M. Deshpande, T.S. Jayram, R. Ramakrishnan, S. Vaithyanathan, in *VLDB '05: Proceedings of the 31st international conference on Very large data bases* (VLDB Endowment, 2005), pp. 970–981
53. D. Burdick, P.M. Deshpande, T.S. Jayram, R. Ramakrishnan, S. Vaithyanathan, in *VLDB* (2006), pp. 391–402
54. D. Burdick, A. Doan, R. Ramakrishnan, S. Vaithyanathan, in *VLDB '07: Proceedings of the 33rd international conference on Very large data bases* (VLDB Endowment, 2007), pp. 39–50
55. T.S. Jayram, A. McGregor, S. Muthukrishnan, E. Vee, ACM Trans. Database Syst. **33**(4), 1 (2008). DOI http://doi.acm.org/10.1145/1412331.1412338
56. R. Cheng, S. Singh, S. Prabhakar, R. Shah, J.S. Vitter, Y. Xia, in *CIKM '06: Proceedings of the 15th ACM international conference on Information and knowledge management* (ACM, New York, NY, USA, 2006), pp. 738–747. DOI http://doi.acm.org/10.1145/1183614.1183719
57. P. Agrawal, J. Widom, Technical report, Stanford University CA, USA (2007)
58. B. Kimelfeld, Y. Sagiv, in *PODS '07: Proceedings of the twenty-sixth ACM SIGMOD-SIGACT-SIGART symposium on Principles of database systems* (ACM, New York, NY, USA, 2007), pp. 303–312. DOI http://doi.acm.org/10.1145/1265530.1265572
59. Y. Tao, X. Xiao, R. Cheng, ACM Trans. Database Syst. **32**(3), 15 (2007). DOI http://doi.acm.org/10.1145/1272743.1272745
60. V. Ljosa, A.K. Singh, in *ICDE* (2007), pp. 946–955
61. C. Böhm, A. Pryakhin, M. Schubert, in *ICDE* (2006), p. 9
62. S. Singh, C. Mayfield, S. Prabhakar, R. Shah, S. Hambrusch, in *Proceedings of the 2007 IEEE 23rd International Conference on Data Engineering (ICDE'07)* (2007)
63. D. Xin, H. Cheng, X. Yan, J. Han, in *KDD* (2006), pp. 444–453
64. D. Xin, J. Han, H. Cheng, X. Li, in *VLDB* (2006), pp. 463–475
65. B. Babcock, C. Olston, in *SIGMOD Conference* (2003), pp. 28–39
66. S. Michel, P. Triantafillou, G. Weikum, in *VLDB* (2005), pp. 637–648
67. D. Dubois, H. Prade, J. Rossazza, International Journal of Intelligent Systems **6**, 167 (1991)
68. V. Tamma, T. Bench-Capon, Knowledge Engineering Review **17**(1), 41 (2002). DOI http://dx.doi.org/10.1017/S0269888902000371
69. E. Rosch, Cognitive Development and Acquisition of Language pp. 111–144 (1973)
70. L.W. Barsalou, Concepts and conceptual development pp. 101–140 (1987)
71. R.M. Nosofsky, Journal of Experimental Psychology: Learning, Memory, and Cognition **14**(1), 54 (1988)
72. E. Rosch, Journal of Experimental Psychology: General **104**, 192 (1975)
73. L.R. Brooks, in *Cognition and categorization*, ed. by E.H. Rosch, B.B. Lloyd (Hillsdale, New York, NY, 1973), pp. 169–211
74. B. Cohen, G.L. Murphy, Cognitive Science **8**, 27 (1984)
75. S. Reed, *Cognition: Theory and Applications* (Wadsworth Publishing. 6 edition, 2003)
76. C.M. Au Yeung, H.F. Leung, Comput. J. (To appear)
77. P. Bose, A. Maheshwari, P. Morin, Computational Geometry: Theory and Applications **24**(3), 135 (2003)
78. D. Cantone, G. Cincotti, A. Ferro, A. Pulvirenti, SIAM Journal on Optimization **16**(2), 434 (2005)
79. P. Indyk, in *Proceedings of the thirty-first annual ACM symposium on Theory of computing* (ACM Press, 1999), pp. 428 – 434
80. S. Bespamyatnikh, K. Kedem, M. Segal, in *Proceedings of the 6th International Workshop on Algorithms and Data Structures* (Springer-Verlag, 1999), pp. 318 – 329
81. R. Xu, D.W. II, Neural Networks, IEEE Transactions on **16**(3), 645 (2005)
82. J.A. Hartigan, M.A. Wong, Applied Statistics **28**, 100 (1979)
83. L. Kaufmann, P.J. Rousseeuw, in *Statistical Data Analysis based on the L1 Norm*, ed. by Y. Dodge (Elsevier/North Holland, Amsterdam, 1987), pp. 405–416
84. R.T. Ng, J. Han, IEEE Trans. on Knowl. and Data Eng. **14**(5), 1003 (2002). DOI http://dx.doi.org/10.1109/TKDE.2002.1033770

85. T. Zhang, R. Ramakrishnan, M. Livny, in *Proceedings of the 1996 ACM SIGMOD International Conference on Management of Data, Montreal, Quebec, Canada, June 4-6, 1996*, ed. by H.V. Jagadish, I.S. Mumick (ACM Press, 1996), pp. 103–114
86. S. Guha, R. Rastogi, K. Shim, SIGMOD Rec. **27**(2), 73 (1998). DOI http://doi.acm.org/10.1145/276305.276312
87. G. Karypis, E.H.S. Han, V. Kumar, Computer **32**(8), 68 (1999). DOI http://dx.doi.org/10.1109/2.781637
88. M. Ester, H.P. Kriegel, J. Sander, X. Xu, in *Second International Conference on Knowledge Discovery and Data Mining*, ed. by E. Simoudis, J. Han, U.M. Fayyad (AAAI Press, 1996), pp. 226–231
89. M. Ankerst, M.M. Breunig, H.P. Kriegel, J. Sander, in *SIGMOD 1999, Proceedings ACM SIGMOD International Conference on Management of Data, June 1-3, 1999, Philadelphia, Pennsylvania, USA*, ed. by A. Delis, C. Faloutsos, S. Ghandeharizadeh (ACM Press, 1999), pp. 49–60
90. A. Hinneburg, D.A. Keim, in *Knowledge Discovery and Data Mining* (1998), pp. 58–65. URL `citeseer.ist.psu.edu/hinneburg98efficient.html`
91. W. Wang, J. Yang, R.R. Muntz, in *VLDB '97: Proceedings of the 23rd International Conference on Very Large Data Bases* (Morgan Kaufmann Publishers Inc., San Francisco, CA, USA, 1997), pp. 186–195
92. R. Agrawal, J. Gehrke, D. Gunopulos, P. Raghavan, in *SIGMOD 1998, Proceedings ACM SIGMOD International Conference on Management of Data, June 2-4, 1998, Seattle, Washington, USA*, ed. by L.M. Haas, A. Tiwary (ACM Press, 1998), pp. 94–105
93. N.A. Campbell, Multivariate Statistical Methods in Physical Anthropology pp. 177–192 (1984)
94. G.M. Foody, N.A. Campbell, N.M. Trodd, T.F. Wood, Photogrammetric Engineering and Remote Sensing **58**, 1335C1341 (1992)
95. E. Cohen, H. Kaplan, in *SIGMOD '04: Proceedings of the 2004 ACM SIGMOD international conference on Management of data* (ACM Press, New York, NY, USA, 2004), pp. 707–718. DOI http://doi.acm.org/10.1145/1007568.1007647
96. E. Cohen, H. Kaplan, J. Comput. Syst. Sci. **73**(3), 265 (2007). DOI http://dx.doi.org/10.1016/j.jcss.2006.10.016
97. X. Lian, L. Chen, in *Proc. 2008 International Conference on Extended Data Base Technology (EDBT'08)* (2008)
98. K. Yi, F. Li, D. Srivastava, G. Kollios, in *Proc. 2008 International Conference on Data Engineering (ICDE'08)* (2008)
99. G. Cormode, F. Li, K. Yi, in *ICDE* (2009), pp. 305–316
100. F. Li, K. Yi, J. Jestes, in *SIGMOD Conference* (2009)
101. J. Li, B. Saha, A. Deshpande, in *VLDB '09: Proceedings of the 35th international conference on Very large data bases* (VLDB Endowment, 2009)
102. Y.H. Wang, Statistica Sinica **3**, 295 (1993)
103. K. Lange, *Numerical analysis for statisticians*. Statistics and computing (1999)
104. G. Cormode, M.N. Garofalakis, in *SIGMOD Conference* (2007), pp. 281–292
105. T.S. Jayram, S. Kale, E. Vee, in *SODA '07: Proceedings of the eighteenth annual ACM-SIAM symposium on Discrete algorithms* (Society for Industrial and Applied Mathematics, Philadelphia, PA, USA, 2007), pp. 346–355
106. T.S. Jayram, A. McGregor, S. Muthukrishnan, E. Vee, in *PODS* (2007), pp. 243–252
107. G. Cormode, F. Korn, S. Muthukrishnan, D. Srivastava, in *PODS'06* (2006)
108. A. Gupta, F.X. Zane, in *SODA '03: Proceedings of the fourteenth annual ACM-SIAM symposium on Discrete algorithms* (Society for Industrial and Applied Mathematics, Philadelphia, PA, USA, 2003), pp. 253–254
109. Y. Zhang, X. Lin, J. Xu, F. Korn, W. Wang, in *ICDE '06: Proceedings of the 22nd International Conference on Data Engineering (ICDE'06)* (IEEE Computer Society, Washington, DC, USA, 2006), p. 51. DOI http://dx.doi.org/10.1109/ICDE.2006.145
110. A. Arasu, G.S. Manku, in *Proceedings of the Twenty-third ACM SIGACT-SIGMOD-SIGART Symposium on Principles of Database Systems (PODS'04)* (Paris, France, 2004)

111. A.C. Gilbert, Y. Kotidis, S. Muthukrishnan, M. Strauss, in *VLDB* (2002), pp. 454–465
112. M. Greenwald, S. Khanna, in *SIGMOD Conference* (2001), pp. 58–66
113. G.S. Manku, S. Rajagopalan, B.G. Lindsay, in *SIGMOD '99: Proceedings of the 1999 ACM SIGMOD international conference on Management of data* (ACM, New York, NY, USA, 1999), pp. 251–262. DOI http://doi.acm.org/10.1145/304182.304204
114. G. Cormode, F. Korn, S. Muthukrishnan, D. Srivastava, in *ICDE'05*
115. G. Das, D. Gunopulos, N. Koudas, N. Sarkas, in *VLDB* (2007), pp. 183–194
116. A. Deshpande, C. Guestrin, S. Madden, J.M. Hellerstein, W. Hong, in *VLDB* (2004), pp. 588–599
117. Z. Liu, K.C. Sia, J. Cho, in *SAC* (2005), pp. 634–641
118. S. Han, E. Chan, R. Cheng, K. yiu Lam, Real-Time Systems **35**(1), 33 (2007)
119. L.J. Roos, A. Wajda, J. Nicol, Comput Biol Med. **16**, 45 (1986)
120. I.P. Fellegi, A.B. Sunter, Journal of the American Statistical Association **64**, 1183 (1969)
121. T.N. Herzog, F.J. Scheuren, W.E. Winkler, in *Data Quality and Record Linkage Techniques* (Springer New York, 2007), pp. 93–106
122. W. Winkler. Matching and record linkage (1995). URL `http://citeseer.ist.psu.edu/winkler95matching.html`
123. W.E. Winkler, in *Proceedings of the Section on Survey Research Methods, American Statistical Association* (1993), pp. 274–279
124. V.I. Levenshtein, Binary codes capable of correcting deletions, insertions, and reversals. Tech. Rep. 8 (1966)
125. D. Gusfield, *Algorithms on strings, trees, and sequences: computer science and computational biology* (Cambridge University Press, New York, NY, USA, 1997)
126. L. Gravano, P.G. Ipeirotis, H.V. Jagadish, N. Koudas, S. Muthukrishnan, D. Srivastava, in *VLDB '01: Proceedings of the 27th International Conference on Very Large Data Bases* (Morgan Kaufmann Publishers Inc., San Francisco, CA, USA, 2001), pp. 491–500
127. S. Chaudhuri, K. Ganjam, V. Ganti, R. Motwani, in *SIGMOD '03: Proceedings of the 2003 ACM SIGMOD international conference on Management of data* (ACM, New York, NY, USA, 2003), pp. 313–324. DOI http://doi.acm.org/10.1145/872757.872796
128. W. Cohen, P. Ravikumar, S. Fienberg. A comparison of string distance metrics for name-matching tasks (2003). URL `http://citeseer.ist.psu.edu/cohen03comparison.html`
129. A. Inan, M. Kantarcioglu, E. Bertino, M. Scannapieco, in *ICDE* (2008), pp. 496–505
130. P. Christen, in *KDD* (2008), pp. 151–159
131. M. Michelson, C.A. Knoblock, in *AAAI* (2006)
132. C. Li, L. Jin, S. Mehrotra, World Wide Web **9**(4), 557 (2006)
133. F.V. Jensen, *An Introduction to Bayesian Networks*, 1st edn. (UCL Press, 1996)
134. R. Kindermann, J. Snell, *Markov random fields and their applications* (American Mathematical Society, Providence, Rhode Island, 1980)
135. E. Castillo, J.M. Gurtiérrez, A.S. Hadi, *Expert Systems and Probabilistic Network Models*. Monographs in Computer Science (1997)
136. H. Frank, Operations Research **17**(4), 583 (1969)
137. R.P. Loui, Commun. ACM **26**(9), 670 (1983). DOI http://doi.acm.org/10.1145/358172.358406
138. R. Hassin, E. Zemel, Mathematics of Operations Research **10**(4), 557 (1985)
139. S.K. Walley, H.H. Tan, in *COCOON '95: Proceedings of the First Annual International Conference on Computing and Combinatorics* (Springer-Verlag, London, UK, 1995), pp. 213–222
140. D. Blai, L. Kaelbling, in *IJCAI Workshop on Adaptive Spatial Representations of Dynamic Environments* (1999)
141. D.D.M.L. Rasteiro, A.J.B. Anjo, Journal of Mathematical Sciences **120**(1), 974 (2004)
142. Fan, Yueyue, Nie, Yu, Networks and Spatial Economics **6**(3-4), 333 (2006). DOI http://dx.doi.org/10.1007/s11067-006-9287-6. URL `http://dx.doi.org/10.1007/s11067-006-9287-6`

143. E. Nikolova, J.A. Kelner, M. Brand, M. Mitzenmacher, in *ESA'06: Proceedings of the 14th conference on Annual European Symposium* (Springer-Verlag, London, UK, 2006), pp. 552–563. DOI http://dx.doi.org/10.1007/11841036_50
144. Y.Y. Fan, R.E. Kalaba, I. J. E. Moore, Journal of Optimization Theory and Applications **127**(3), 497 (2005)
145. J. Ghosh, H.Q. Ngo, S. Yoon, C. Qiao, in *INFOCOM* (2007), pp. 1721–1729
146. A. Chang, E. Amir, in *23rd Conference on Uncertainty in Artificial Intelligence (UAI'07)* (2007)
147. A.B. Kahng, M. Mani, A. Zelikovsky, G. Bhatia, Traversing probabilistic graphs. Tech. Rep. 990010 (1999). URL `citeseer.ist.psu.edu/mani99traversing.html`
148. J.S. Provan, M.O. Ball., SIAM Journal on Computing **12**(4), 777 (1983)
149. L. Fu, D. Sun, L.R. Rilett, Computers and Operations Research **33**(11), 3324 (2006)
150. P. Hart, N. Nilsson, B. Raphael, Systems Science and Cybernetics, IEEE Transactions on **4**(2), 100 (July 1968). DOI 10.1109/TSSC.1968.300136
151. N.J. Nilsson, *Problem-Solving Methods in Artificial Intelligence* (McGraw-Hill Pub. Co., 1971)
152. P. Sanders, D. Schultes, in *ESA* (2005), pp. 568–579
153. P. Sanders, D. Schultes, in *ESA* (2006), pp. 804–816
154. G. Ertl, OR Spektrum **20**, 15 (1998)
155. G.G. Filho, H. Samet, A linear iterative approach for hierarchical shortest path finding. Tech. rep., Computer Science Department CS-TR-4417, University of Maryland
156. H. Bast, S. Funke, D. Matijevic, in *9th DIMACS Implementation Challenge — Shortest Path*, ed. by C. Demetrescu, A. Goldberg, D. Johnson (DIMACS, Piscataway, New Jersey, 2006)
157. H. Bast, S. Funke, D. Matijevic, P. Sanders, D. Schultes, in *ALENEX* (2007)
158. H. Gonzalez, J. Han, X. Li, M. Myslinska, J.P. Sondag, in *VLDB '07: Proceedings of the 33rd international conference on Very large data bases* (VLDB Endowment, 2007), pp. 794–805
159. A.B. Kurzhanski, P. Varaiya, SIAM J. Control Optim. **41**(1), 181 (2002). DOI http://dx.doi.org/10.1137/S0363012999361093
160. S. Baswana, S. Sen, in *FST TCS 2000: Proceedings of the 20th Conference on Foundations of Software Technology and Theoretical Computer Science* (Springer-Verlag, London, UK, 2000), pp. 252–263
161. Y. Wu, J. Xu, Y. Hu, Q. Yang, in *Intelligent Transportation Systems* (2003), pp. 1511 – 1514
162. E.P.F. Chan, J. Zhang, in *CIKM '07: Proceedings of the sixteenth ACM conference on Conference on information and knowledge management* (ACM, New York, NY, USA, 2007), pp. 371–380. DOI http://doi.acm.org/10.1145/1321440.1321494
163. H. Samet, J. Sankaranarayanan, H. Alborzi, in *SIGMOD '08: Proceedings of the 2008 ACM SIGMOD international conference on Management of data* (ACM, New York, NY, USA, 2008), pp. 43–54. DOI http://doi.acm.org/10.1145/1376616.1376623
164. L. Devroye, G. Lugosi, *Combinatorial Methods in Density Estimation* (Springer; 1 edition, 2001)
165. L. Devroye, *A course in density estimation* (Birkhauser Boston Inc, 1987)
166. Y. Kanazawa, Annals of statistics **20**(1), 291 (1992)
167. I.S. Abramson, Annals of statistics **10**(4), 1217 (1982)
168. L. Breiman, W. Meisel, E. Purcell, Technometrics **19**(2), 135 (1977)
169. Y. Mack, M. Rosenblatt, Journal of Multivariate Analysis **9**, 1 (1979)
170. D. Gunopoulos, G. Kollios, V. Tsotras, C. Domeniconi, VLDB Journal **14**(2), 137 (2005)
171. B.W. Silverman, *Density Estimation for Statistics and Data Analysis (Hardcover)* (Chapman and Hall, 1986)
172. M. Rudemo, Scandinavian Journal of Statistics **9**, 65 (1982)
173. A.W. Bowman, Biometrika **71**(2), 353 (1984)
174. S.R. Sain, K.A. Baggerly, D.W. Scott, Journal of the American Statistical Association **89**(427), 807 (1994). URL `citeseer.ist.psu.edu/sain92crossvalidation.html`
175. D. Scott, S. Sain, Handbook of Statistics **23: Data Mining and Computational Statistics** (2004)

176. V. Hodge, J. Austin, Artif. Intell. Rev. **22**(2), 85 (2004). DOI http://dx.doi.org/10.1023/B:AIRE.0000045502.10941.a9
177. S. Walfish, Pharmaceutical Technology **30**(11), 82 (2006)
178. M. Tarter, Comput Programs Biomed. **10**(1), 55 (1979)
179. P.N. Yianilos, in *Proceedings of the fourth annual ACM-SIAM Symposium on Discrete algorithms* (Society for Industrial and Applied Mathematics, 1993), pp. 311 – 321
180. T. Bozkaya, M. Ozsoyoglu, ACM Trans. Database Syst. **24**(3), 361 (1999). DOI http://doi.acm.org/10.1145/328939.328959
181. D. Angluin, L.G. Valiant, in *STOC '77: Proceedings of the ninth annual ACM symposium on Theory of computing* (ACM Press, New York, NY, USA, 1977), pp. 30–41. DOI http://doi.acm.org/10.1145/800105.803393
182. D. Angluin, L.G. Valiant, in *Proceedings of the ninth annual ACM symposium on Theory of computing (STOC'77)* (ACM Press, New York, NY, USA, 1977), pp. 30–41. DOI http://doi.acm.org/10.1145/800105.803393
183. W. Hoeffding, Annals of Mathematical Statistics **27**, 713 (1956)
184. R. Motwani, P. Raghavan, *Randomized Algorithms* (Cambridge University Press, United Kingdom, 1995)
185. J.L. Hodges, J.L.L. Cam, The Annals of Mathematical Statistics **31**(3), 737 (1960)
186. L.L. Cam, Pacific J. Math. **10**, 1181 (1960)
187. J.I. Munro, M. Paterson, Theor. Comput. Sci. **12**, 315 (1980)
188. X. Lin, H. Lu, J. Xu, J.X. Yu, in *ICDE* (2004), pp. 362–374
189. G.S. Manku, S. Rajagopalan, B.G. Lindsay, in *SIGMOD Conference* (1998), pp. 426–435
190. J.R. Evans, E. Minieka, *Optimization Algorithms for Networks and Graphs* (Marcel Dekker Inc, 1992)
191. R. Righter, in *Stochastic Orders and Their Applications*, ed. by M. Shaked, J. Shanthikumar (Academic Press, 1994), pp. 381–432
192. B.W. Kernighan, S. Lin, The Bell System Technical Journal **49**(2), 291 (1970)
193. M.T. Goodrich, J. Comput. Syst. Sci. **51**(3), 374 (1995). DOI http://dx.doi.org/10.1006/jcss.1995.1076
194. R.R. He, H.X. Liu, A.L. Kornhauser, B. Ran, Center for Traffic Simulation Studies (Paper UCI-ITS-TS-02-14)
195. A. Polus, Transportation **8**(2), 141 (1979)
196. Y. Tao, D. Papadias, IEEE Trans. on Knowl. and Data Eng. **18**(3), 377 (2006). DOI http://dx.doi.org/10.1109/TKDE.2006.48
197. M. Lacroix, P. Lavency, in *VLDB '87: Proceedings of the 13th International Conference on Very Large Data Bases* (Morgan Kaufmann Publishers Inc., San Francisco, CA, USA, 1987), pp. 217–225
198. D. Xin, J. Han, in *ICDE* (2008), pp. 1092–1100
199. W. Kießling, G. Köstler, in *VLDB* (2002), pp. 990–1001
200. K. Govindarajan, B. Jayaraman, S. Mantha, New Gen. Comput. **19**(1), 57 (2001). DOI http://dx.doi.org/10.1007/BF03037534
201. J. Chomicki, in *EDBT '02: Proceedings of the 8th International Conference on Extending Database Technology* (Springer-Verlag, London, UK, 2002), pp. 34–51